UNIVERSITY OF
WOLVERHAMPTON
KNOWLEDGE • INNOVATION • ENTERPRISE

P   CELL CULTURE

The INTRODUCTION TO BIOTECHNIQUES series

Editors:

David Rickwood, Department of Biological and Chemical Sciences, University of Essex, Colchester, Essex

Chris Howe, Department of Biochemistry, University of Cambridge, Cambridge

CENTRIFUGATION
RADIOISOTOPES
LIGHT MICROSCOPY
ANIMAL CELL CULTURE
GEL ELECTROPHORESIS: PROTEINS
PCR, SECOND EDITION
MICROBIAL CULTURE
ANTIBODY TECHNOLOGY
GENE TECHNOLOGY
LIPID ANALYSIS
GEL ELECTROPHORESIS: NUCLEIC ACIDS
LIGHT SPECTROSCOPY
DNA SEQUENCING
MEMBRANE ANALYSIS
PLANT CELL CULTURE

Forthcoming title

NUCLEIC ACID HYBRIDIZATION

PLANT CELL CULTURE

Hamish A. Collin

The University of Liverpool, School of Biological Sciences,
Donnan Laboratories, Liverpool L69 3BX, UK

Sue Edwards

Unit of Vegetation Sciences, Institute of Environmental and
Biological Sciences, Lancaster University, Lancaster LA1 4YQ,
UK

βIOS
SCIENTIFIC
PUBLISHERS

© BIOS Scientific Publishers Limited, 1998

First published 1998

A CIP catalogue record for this book is available from the British Library.

ISBN 1 872748 47 3

BIOS Scientific Publishers Ltd
9 Newtec Place, Magdalen Road, Oxford OX4 1RE, UK
Tel. +44 (0) 1865 726286. Fax +44 (0) 1865 246823
World Wide Web home page: http://www.bios.co.uk/

DISTRIBUTORS

Australia and New Zealand
Blackwell Science Asia
54 University Street
Carlton
South Victoria 3053

India
Viva Books Private Limited
4325/3 Ansari Road
Daryaganj
New Delhi 110002

Published in the United States of America, its dependent territories and Canada by Springer-Verlag New York Inc., 175 Fifth Avenue, New York, NY 10010-7858, in association with BIOS Scientific Publishers Ltd.

Published in Hong Kong, Taiwan, Singapore, Thailand, Cambodia, Korea, The Philippines, Indonesia, The People's Republic of China, Brunei, Laos, Malaysia, Macau and Vietnam by Springer-Verlag Singapore Pte Ltd, 1 Tannery Road, Singapore 347719, in association with BIOS Scientific Publishers Ltd.

Production Editor: Andrea Bosher.
Typeset by Chandos Electronic Publishing, Stanton Harcourt, UK.
Printed by Biddles Ltd, Guildford, UK.

Contents

Abbreviations

2,4-D	2,4-Dichlorophenoxyacetic acid
ABA	abscisic acid
BAP	6-benzylaminopurine
DMSO	dimethyl sulphoxide
EDTA	ethylene diamine tetra-acetic acid
EPSPS	5-enolpyruvyl-3-phosphate synthase
FDA	fluorescein diacetate
GA3	gibberellic acid
GB5	Gamborg B5
H6H	hyoscyamine 6 β-hydroxylase
IAA	indole acetic acid
IBA	indole-3-butyric acid
K	kinetin
Lea	late embryogenesis abundant (genes)
MS	Murashige and Skoog
NAA	1-Naphthalene acetic acid
NOA	2-naphthoxyacetic acid
PAL	phenylammonia lyase
pCPA	*p*-Chlorophenoxyacetic acid
PCV	packed cell volume
PEG	polyethylene glycol
RAI	radioactive immunoassay
RAPD	random amplified polymorphic DNA
RFLP	restriction fragment length polymorphism
SDS	sodium dodecylsulphate
SH	Schenk and Hildebrandt
TDC	tryptophan decarboxylase
Zea	zeatin
ZiP	N-isoperitenylamino acid

Preface

Plant tissue culture went through a phase when it was regarded as a subject in itself. There was enormous excitement as each new discovery was described to packed audiences in conference halls, then the subject became a technique as science moved on. Plant tissue culture took its place as a key component in plant breeding, plant propagation and biotechnology. Now haploid culture is routine for some crops, helping to shorten the production of new varieties by years. Micropropagation is part of a major commercial enterprise which is producing attractive house plants as well as the multiplication of elite food and timber-producing trees. Genetically engineered crops are in the supermarkets and seem to be here to stay, while the sophisticated secondary metabolism of plant cells is being unravelled and put to use by the industrial chemist to help synthesize pharmaceuticals. In all of these successful applications there is a need to know how to isolate suitable plant tissue as a starting point, to maintain the tissue under *in vitro* conditions in an undifferentiated or partially differentiated form, and finally to redifferentiate the tissue back to the intact plant. These basic steps are described in sufficient detail in this volume to enable anyone new to the technique to start, maintain and redifferentiate a tissue culture. Also described, in a carefully structured format, are details of how the techniques can be applied to haploid and protoplast culture, the production and selection of somaclonal variation, production and pitfalls in micropropagation, and the possibilities of genetic manipulation. The description of the theory and practice should provide the novice with sufficient background for a start to be made on all of these techniques.

Finally I would like to acknowledge the contribution made to this book by the discussions with and results of all those research students whom I have supervised in the past on tissue culture-related projects. I would also like to acknowledge my wife for her support and members of staff of the Genome Structure and Function Research Group within the School of Biological Sciences for their patience in the preparation of this book.

Hamish A. Collin

1 Introduction

The technique of plant cell tissue culture occupies a key role in the 'second green revolution' in which gene modification and biotechnology are being used to improve crop yield and quality. Those working with plant cell tissue culture are still concerned with the details of the technique but are now more involved in its application to fundamental aspects of plant cell differentiation and development and to the problems of crop improvement. To appreciate the current achievements of plant cell tissue culture a short history of its development is described below.

1.1 Historical perspective

It was Schwann who first drew attention to the fact that an individual cell has the ability to both grow and divide in a self regulatory fashion and that an individual cell is also totipotent, that is, whatever its level of differentiation it will retain the capacity to regenerate into the whole organism of which it was once part. These ideas are encapsulated in the 'Cell Theory' and are a fundamental basis underlying plant cell tissue culture today.

The subsequent interest in establishing the functional relationships between tissues and of investigating the possibility for cells to develop in isolation from the intact plant gave rise at the beginning of this century to the techniques of plant cell tissue culture. Gottlieb Haberlandt in 1902 published the first paper on plant cell tissue culture entitled, 'Experiments on the culture of isolated plant cells'. He worked with individual cells isolated from the palisade layer of leaf tissue, with parenchyma cells, epidermal cells and hairs of a number of monocotyledonous species. Although he was able to maintain the cells alive he was not able to induce them to divide. Little further progress was made until White [1] cultured tomato roots on a simple medium containing inorganic salts, sucrose and yeast extract, which we now know to be rich in B vitamins. One of the reasons for the lack of success of Haberlandt and other earlier workers in stimulating division in isolated cells was the fact that they employed a more simple medium which did not contain growth regulators. Organised tissue, such as excised roots, appeared capable of synthesis of a variety of organic components, including plant growth regulators such as auxins and cytokinins, which have been found to be crucial to the initiation and maintenance of cell division in isolated cells.

Over the next 5 years the recognition of the importance of B vitamins in yeast extract and the auxin, indole-3-acetic acid (IAA) allowed significant advances to be made. Gautheret found that cambial tissue of *Salix caprea* and *Populus alba* could proliferate and divide for several months after aseptic isolation on a simple medium but growth was limited. In 1939, however, Gautheret [2] reported the propagation of carrot as the first plant tissue culture of unlimited growth. By including IAA as the auxin source in the media he was able to stimulate the growth of undifferentiated tissue on cut surfaces of sterile explants. This tissue, termed callus, was similar in appearance to wound tissue and subsequently it was found that callus could be subcultured indefinitely.

The production of a callus from explant tissue is accompanied by a series of changes in both the appearance and metabolism of the cells, although the precise nature of the response to culture conditions is dependent upon the medium and explant used. When a surface is cut, the cells at the wound site undergo division, with the first divisions being synchronous. Direct comparisons can be made here with the formation of a wound cambium, for in both systems the cells undergo division but do not subsequently expand in size. After 4–6 weeks of culture and continuous cell division, the explant tissue may have produced double its original weight in callus tissue. If the callus tissue is excised and placed on fresh nutrient media it will continue to proliferate (*Figure 1.1*). The production of callus is considered to be a process of de-differentiation of organised tissue, as indicated by changes in morphology and metabolic activity. It is possible therefore that by manipulating components of the media, the process of de-differentiation can be reversed and the tissue re-differentiated back to an organised form (*Figures 1.2* and *1.3*).

This process of de-differentiation and re-differentiation has now been performed with many species. Control of the process of differentiation was found to be dependent upon the presence of auxin and cytokinin, which are two classes of plant growth regulators. In a classic experiment in 1957, Skoog and Miller [3] demonstrated that root and shoot formation from undifferentiated callus of tobacco was dependent upon the balance of auxin and cytokinin and subsequently this was

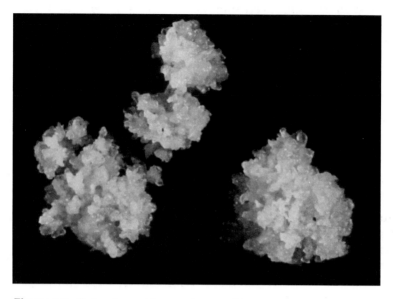

Figure 1.1. Callus formed from onion seedlings.

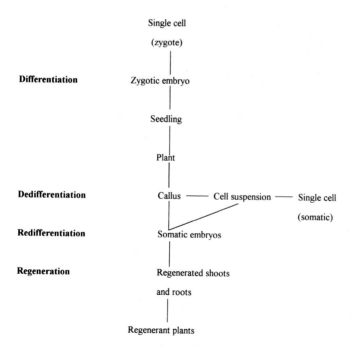

Figure 1.2. Sequence of dedifferentiation and differentiation in plant tissue cultures.

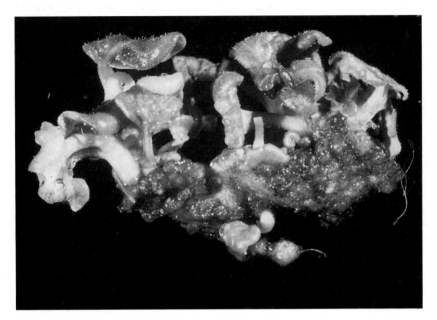

Figure 1.3. Re-differentiation of callus of *Coleus blumei* to give shoots only.

found to be true for other species (*Figure 1.3*). Shortly afterwards Steward *et al.* [4] showed that in the presence of a nutrient medium and a complex additive such as coconut milk, carrot callus tissue regenerated along a specifically embryogenic pathway to form embryos which could develop into intact plants. Coconut milk is the endosperm in the seed husk and, as such, is rich in plant growth regulators. Coconut milk has now been replaced by a mixture of auxins and cytokinins to initiate embryo formation in a large number of species (*Figure 1.4*).

Vasil and Hildebrandt [5] were the first to prove the early cell theory of Schwann by regenerating a tobacco plant from a single cell derived from a suspension of cultured cells. Growth of a single cell in a microchamber to give rise to a callus was then regenerated. Thus the principles of tissue culture were established over a long period of development (>100 years) with the most rapid progress in the last 30–40 years.

Since the earlier studies, the nutritional requirements for the tissue culture of a large number of species have been established and much work has been done on characterizing the biochemistry, physiology and cytology of cultured cells. For a more detailed history on the origins of tissue culture, see Gautheret [2].

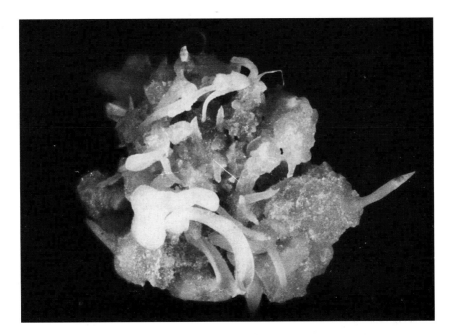

Figure 1.4. Differentiation into embryos on celery callus.

1.2 Advantages of tissue culture over intact plants

The original technique of callus initiation and regeneration has now been expanded to include a whole new range of techniques that have a direct commercial application as well as enabling a study of basic cell genetics and biochemistry. These techniques include the large-scale culture of plant cells, anther, ovule and embryo culture, protoplast isolation and fusion, cell selection, meristem and bud culture and genetic transformation of cells and plants. The contribution of these techniques are that:

1. The biochemical engineer can grow plant cells in liquid culture on a large scale so that they provide potentially a more convenient and profitable source of plant secondary products than the intact plants.

2. The production of dihaploid plants from haploid cultures shortens the time taken to achieve uniform homozygous lines and varieties.

3. The crossing of distantly related species by protoplast isolation and somatic fusion increases the possibility for the transfer and expression of novel variation in domestic crops.

4. Cell selection increases the potential number of individuals in a screening program so that, in an *in vitro* or cell screen for salt tolerance or herbicide resistance, millions of potential and variable individuals can be screened in a limited space very rapidly.

5. Micropropagation using meristem and shoot culture techniques allows the production of large numbers of uniform individuals of horticultural species from limited starting material.

6. Genetic transformation of cells enables very specific information to be introduced into single cells which can then be regenerated. In this way there is no major disturbance to the maternal genome, unlike the more conventional crossing of two individuals.

There are numerous very detailed variants of each of these techniques. It was not the intention of this book to reproduce this detail, but to present an overview, which contains sufficient practical information to carry out any one of these techniques. The book contains a description of the laboratory facilities required for tissue culture work and the aseptic techniques, media and media preparation necessary for the different types of cultures, cell suspension, callus, embryos, haploids and protoplasts. The application of these techniques to crop improvement and secondary product formation is also presented.

References

1. **White, P.R.** (1934) *Plant Physiol.* **9,** 585–600.
2. **Gautheret, R.J.** (1985) In *Cell Culture and Somatic Cell Genetics of Plants* (I. Vasil, ed.), Vol. 2. Academic Press, London, pp. 1–59.
3. **Skoog, F. and Miller, W.** (1957) In *Biological Action of Growth Substances* (H.K. Porter, ed.). Symposium of Society for Experimental Biology, Cambridge University Press, Cambridge, **11,** 118–131.
4. **Steward, F.C., Mapes, M.O. and Mears, K.** (1958) *Am. J. Bot.* **45,** 705–708.
5. **Vasil, V. and Hildebrandt, A.C.** (1965) *Science,* **150,** 889–892.

2 Equipment and general practise

This chapter provides basic advice of what equipment is required when setting up a plant cell tissue culture laboratory. The main advantage of using plant cell tissue cultures, rather than intact plants is that control of both the physical and chemical environment is more easily achieved. In order to attain this control, however, aseptic conditions must be established and maintained. Plant cell culture media is a rich source of both organic (e.g. sucrose 1–5% w/v) and inorganic compounds and as such is a good growth medium for both fungal and bacterial micro-organisms as well as isolated plant cells. Micro-organisms tend to have faster growth rates than plant cells and if aseptic conditions are not maintained, the growth media can quickly become contaminated. Micro-organisms growing on any substrate will quickly alter the chemical environment by not only using the growth media and plant cells as food substrates, but also by excreting metabolites into the medium. All three processes will lead to a rapid loss of defined and controled conditions in the culture vessels.

Sources of contamination may come from the explant itself, culture vessels, media, the environment wherever the culture is handled or manipulated and finally from the instruments used in initiating and subculturing the cultures. The application of aseptic techniques is one of the most important practical aspects of plant cell tissue culture. However, before considering the more specific aspects of callus initiation and media preparation, the layout and design of a typical plant cell tissue culture facility is described.

2.1 Conditions for media preparation

Many authors have written about the advantage of having a separate room set aside for the preparation of media, situated away from the general laboratory and which should contain a range of culture vessels, required chemicals (Chapter 4), a balance, pH meter, bunsen burners, utensils for media preparation and ideally a peristaltic pump for dispensing aliquots of media into vessels prior to autoclaving as well as a source of distilled water. This room should be kept clean, tidy and dust-free with surfaces and equipment wiped down regularly and certainly before and after use in order that microbial spore loading in the air is kept to a minimum. Balances and pH meters should be kept scrupulously clean to prevent undesirable and unquantifiable chemical contamination of the media. If media preparation has to be done in the general laboratory then extra care is required to ensure that all equipment to be used is cleaned beforehand.

2.1.1 Types of culture vessels

Plant cell tissue cultures can be grown in plastic and glass vessels that come in a range of shapes and sizes. The shape and size of container depends on the form of the tissue culture, what containers are available, storage systems used and personal taste.   In the simplest terms there are three forms of culture vessel all of which are easily obtainable (see Appendix A). These are listed below:

1.   Petri dishes, both plastic and glass petri dishes (5.0–9.0 cm diameter). The plastic dishes are pre-sterilized whereas the glass dishes must be autoclaved.

2.   Wide-necked Universal vials, both plastic and glass. Plastic vials are pre-sterilized.

3.   Wide-necked conical glass culture flasks (100–2000 ml).

4.   Pre-sterilized plastic containers with fitted lids (Magenta).

2.1.2 Autoclaving media and glassware

All culture media and any equipment or instruments that will be required for culture work must be sterilized in order to remove potential sources of contamination. Sterilization of instruments, culture vessels and media is usually performed by steam (wet) sterilization in an autoclave. This process of sterilization is quicker than dry sterilization in hot-air ovens and is more effective in removing microbial contamination. Autoclaves should not be situated in the culture room since they are a potential source of contamination, but should be placed somewhere close by for convenience.

There are many types of autoclaves and although some are more complex than others, the basic principles of operation remain the same. Once the equipment to be sterilized has been placed inside the autoclave, the door is sealed and steam is pumped inside, either from some external source or more usually by the boiling of water internally within the autoclave. The autoclave is purged of air by the steam, and once this is achieved the temperature, and hence pressure, is increased to a pre-set level and maintained at this level for a set period of time. For most applications such as empty vessels, scalpels and individual vessels containing less than 2 l of liquid this is usually 121°C 15 lb in^{-2} (103.5 kPa) for 15–20 min. Where larger volumes of liquid are being sterilized then the length of time at which the temperature and pressure are maintained may have to be increased to 40 min, but reference must be made to individual autoclave manuals for relevant advice. Prior to autoclaving, all instruments are wrapped in kitchen aluminum foil and open glass and plastic containers are capped with a double layer of aluminum foil. Eight types of plastics that can be repeatedly autoclaved include those made of polypropylene, polymethylpentene, polyallomer, Tefzel, RETFE and Teflon RREP; but for other materials manufacturer's advice must be sought as to the suitability of individual plastics for use in autoclaves. Plastics may take longer to reach sterilization temperatures and therefore the time in the autoclave may have to be extended. The exact duration of the autoclave time should be obtained from the manufacturers. Bottles to be autoclaved should have their caps slightly loosened, other items should not be sealed tightly or sterilization may not be effective. Metal pipette cans are advised for sterilizing glass pipettes.

2.1.3 *Testing for successful autoclaving*

Modern autoclaves have sensors or probes within the pressure vessel which monitor and record the conditions within the autoclave throughout the run and the information is recorded in the form of a print-out. In the absence of such technology other methods can be employed:

1. Autoclave tape, which is initially plain, but on exposure to high temperatures dark stripes are produced across the tape, may be used. This is the most commonly used method, it is reasonably cheap, but it is not the most reliable, since the tape only indicates whether the container has been through a high temperature program and not whether it is sterile.

2. Sterilization tubes, which change color after exposure to appropriate autoclave conditions.

3. Autoclave strips, which turn progressively blue along their length, and show a safe sterilization when the blue reaches the indicated 'safe mark'.

Modern autoclaves have a drying cycle at the end of the sterilization run, and do not allow the autoclave to be opened until it has dropped to room temperature and pressure. With older models, the door can be opened immediately the autoclave run has finished and the pressure has dropped to zero, although the drum inside is hot. Opening the door can lead to a rapid change in temperature resulting in breakage of glassware and possibly contamination, as non-sterile air is drawn into the autoclave. Once the autoclave has been opened, any tops which were loosened prior to autoclaving should be tightened.

2.2 Sterile cabinets or rooms for aseptic transfer

After all media or instruments have been sterilized it becomes important for all further manipulations to be carried out using aseptic techniques (Chapter 3). Conventionally, laminar flow cabinets are employed for this purpose (see Appendix A). Laminar flow cabinets are enclosed on 5 sides and consist of a coarse filter, usually situated on the roof, through which air is drawn to filter off particulate matter and then the air is drawn through a series of bacterial filters forcing sterile air out over the work bench and into the face of the operator. This cabinet, although suitable for plant cell culture, is not suitable for work involving animal cell cultures or with serious pathogens where more protection is required for the operator. Many cabinets are fitted with fluorescent lighting so that the cabinet can be illuminated during operation. Some have UV lamps, which can be left on when the cabinet is not in use and no personnel are nearby, in order to sterilize the cabinet. However, UV lamps must be switched off when the cabinet is in use. If a laminar flow cabinet is not available, a suitable but less efficient alternative is an inoculating room. This is a small room that contains a hard bench surface that can be surface-sterilized and a bunsen burner (Chapter 3) and usually a UV lamp that can be directed over the work surface to sterilize the surface prior to use. Both laminar flow cabinets and inoculating rooms should have adequate lighting and must be situated away from potential sources of drafts.

2.3 Temperature and lighting requirements for growth of cultures

The main physical requirement for the growth and maintenance of plant cell tissue cultures is the ability to maintain a constant temperature of 25°C ± 2°C. Where plant cell tissue cultures are being maintained on a small scale then a standard, microbiological incubator, with lighting if possible, will suffice. If a more ambitious plant cell tissue culture program is undertaken then the setting up of a tissue culture room is recommended. This room should be fitted with a temperature thermostat and shelves on which to place the cultures with the underneath of the shelves being used to support fluorescent strips to illuminate those cultures that require light. The lights should be attached to timers so that the length of light and darkness can be controled. Finally it is recommended that cupboards are fitted for cultures that do not require lights.

For the initiation of suspension cultures, shakers will be required (see Appendix A). The cheapest and certainly the easiest to obtain are open platform orbital shakers that can be situated on the bench, or floor of a tissue culture room. If these are not practical, then enclosed orbital shakers with fluorescent lights can be obtained but the enclosed type are considerably more expensive. When purchasing shakers it is necessary to obtain those with a integral cooling circuit, as the motors can generate substantial amounts of heat which, if not cooled, can be directed out into the culture room and contribute to an increase in the ambient temperature.

3 Aseptic techniques in plant tissue culture

This chapter aims to illustrate the most common aseptic procedures used in plant cell tissue culture.

3.1 Preparation of the laminar flow cabinet

Most aseptic manipulations will take place in a laminar flow cabinet. There are a number of basic procedures that must be followed while working in a cabinet.

1. Before starting any sterile transfer, the operator's hands and lower arms are washed in warm soapy water and either allowed to air dry, or dried with a clean paper towel.

2. All internal surfaces of the cabinet are surface-sterilized by swabbing with 70% ethanol or methylated spirits. The air-flow is switched on to allow equilibration for 10–15 minutes then the cabinet is surface-sterilized again immediately prior to use. The back grill must not be touched as this may introduce contamination onto the back filter.

3. Only essential items are placed in the cabinet since aseptic conditions are maintained more easily in an uncluttered area. For example the bunsen burner is kept near the front; a conical flask containing ethanol for instrument sterilization again near the front, but away from the bunsen burner and those vessels immediately in use for that particular manipulation lining the sides so as to keep the center free.

4. A trolley with shelves is useful for holding stocks of equipment, for example, further vessels, pipettes and appropriate containers for waste disposal and this is kept close at hand.

3.2 Handling pipettes and pipette tips

When distributing aliquots of sterile solution within a flow cabinet or room, sterilized re-usable glass or sterile disposable pipettes are used. In order to sterilize re-usable glass pipettes, they should always be loaded tips first into a can onto a pad of non-absorbent cotton wool, then autoclaved. The pipettes are handled in the laminar flow cabinet as follows.

1. The can is always opened under aseptic conditions by holding the can in a near-horizontal position, angled slightly downwards, and shaking gently to encourage the pipette ends to slide out from the edge of the can. Individual pipettes are removed without touching the exposed end and are attached to fillers. The pipette tip must not touch the cabinet floor. If pipettes are made of glass then it is advisable to pass the tip through a bunsen flame very briefly (<1 s) prior to use. The can lid is replaced after each pipette is removed.

2. When sterile disposable pipettes are being used, the individual packaging at the end of the pipette that is to be attached to the pipette filler is opened, then the filler is attached to the pipette and the remainder of packaging is removed.

3. If the transfer is by Gilson pipette, a sterile tip must be attached. The sterile tips are placed in the transfer cabinet, the cover removed and the pipette pressed down onto an individual tip before withdrawing the tip. The tip is never picked up by hand and pressed onto the pipette as this increases the risk of contamination.

3.3 Aseptic transfer of media

Where large volumes of media are autoclaved, small aliquots are transferred aseptically to smaller containers in the laminar cabinet.

1. The cap of the media container is loosened and held in the curve of the little finger of the left hand, or right hand if the operator is left-handed. This maneuver does require a little practise but it reduces the risk of contamination by preventing contact of the cap with any potential contaminated surfaces and also encourages the

operator to replace the cap whenever the bottle is not in use. The advantage of this method is that it leaves the right hand free for manipulation.

2. The container is held near the base and not the neck, then, with the container in the near horizontal position, the neck of the bottle is flamed. The required volume of media is withdrawn into the pipette, then the media is dispensed into the sterilized culture vessel observing the same aseptic procedures (handling near the base and flaming the neck etc.).

3. An extra precaution is to decant a volume of media from the larger container into a small sterile container then to transfer aliquots from here to the culture vessels. By not inserting the pipette into the main container regularly, the risk of contaminating all the media is reduced. The small vessel can be changed regularly to contain any contamination.

4. Where only small volumes of liquid need to be sterilized, syringe filter units are available commercially. The protective backing is peeled away from the unit, then the syringe is attached while the filter is still in its plastic mold. The filter unit itself must not be handled or contamination will occur.

3.4 Aseptic transfer of callus and cell suspensions

The following sequence should be adopted for transferring callus tissue from one sterile container to another in either a transfer cabinet or room.

1. A small diameter spoon spatula is dipped in 70% ethanol then flamed.

2. The top of container A is removed then the neck flamed while holding near the base.

3. Container A is held in a near horizontal position and the spatula inserted without touching the sides and the callus tissue is excised (*Figure 3.1*).

4. The neck of container A is flamed and the cap replaced.

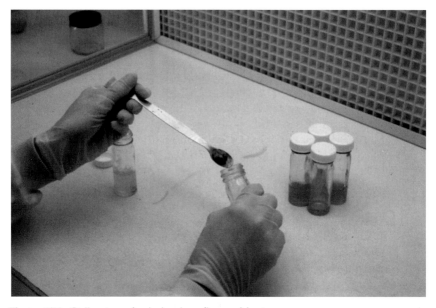

Figure 3.1. Callus transfer in laminar flow cabinet.

5. The cap of container B is removed, the neck flamed and the excised tissue placed in container B.

6. The spatula is withdrawn and placed in 70% ethanol.

7. The neck of container B is flamed and the cap is replaced.

Where pre-sterilized plastic containers are used, the neck is not flamed in the transfer process. It may be easier to withdraw the culture from the container and dissect it on the surface of a sterile petri dish before transferring portions to fresh medium (*Figure 3.2*).

When cell suspensions are transferred to new medium, sterilized glass or pre-sterilized plastic pipettes are used to transfer a specific volume of the suspension. Since the pipettes are already sterile, the above procedure is adopted except for step 1 and the final stage of step 6.

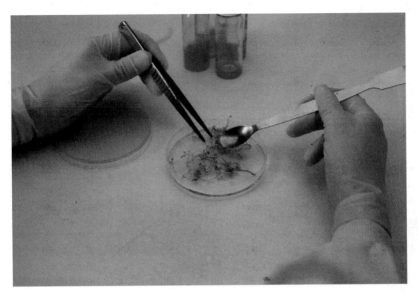

Figure 3.2. Separation of portions of differentiated callus on sterile petri dish prior to transfer to fresh media.

4 Preparation of plant tissue culture media

This chapter provides basic information on the composition and preparation of media suitable for both callus initiation and maintenance and the initiation and growth of cell suspension cultures. More specialized media and decisions on choice of media are given in later chapters.

4.1 Composition of media

There are eight major groups of components in media required for plant cell tissue culture. These are listed below.

4.1.1 Major inorganic nutrients

These are nitrogen (as NO_3 and NH_4), phosphorus (as PO_4), potassium, sulfur (as SO_4), calcium and magnesium, all of which are usually present in mM quantities (see *Table 4.1*). These elements have both structural and functional roles in protein synthesis (particularly N and S), nucleotide synthesis (P, N, S) cell wall synthesis (Ca), enzyme co-factors (Mg) and membrane integrity (Mg).

4.1.2 Micro elements

These are generally supplied in µM quantities (*Table 4.1*) and include manganese, zinc, boron, copper, cobalt and molybdenum. Many of these have important functions in enzyme function as co-factors.

4.1.3 Iron

This is an important enzyme co-factor which is supplied in mM quantities (*Table 4.1*) in a chelated form (e.g. Fe ethyldiamine tetra

Table 4.1.   Composition of commonly used plant tissue culture media

Constituents	Concentration in culture medium (mg l^{-1})			
	Murashige and Skoog (MS)	Schenk and Hildebrandt (SH)	Gamborg B5 (GB5)	Nitsch and Nitsch
Macro nutrients				
KNO_3	1 900	2 500	2 500	950
NH_4NO_3	1 650			720
$NH_4H_2PO_2$		300		
$(NH_4)_2SO_4$			134	
$MgSO_4 \cdot 7H_2O$	370	400	250	185
$CaCl_2 \cdot 2H_2O$	440	200	150	220
KH_2PO_4	170			68
$NaH_2PO_4 \cdot H_2O$			150	
Micro nutrients				
$MnSO_4 \cdot H_2O$		10.0	10.0	
$MnSO_4 4H_2O$	22.3			25
KI	0.83	1.0	0.75	
H_3BO_3	6.2	5.0	3.0	10
$ZnSO_4 \cdot 7H_2O$	8.6	1.0	2.0	10
$CuSO_4 \cdot 5H_2O$	0.025	0.2	0.025	0.025
$Na_2MoO_4 \cdot 2H_2O$	0.25	0.1	0.25	0.25
$CoCl_2 \cdot 6H_2O$	0.025	0.1	0.025	
$FeSO_4 \cdot 7H_2O$	27.8	15.0	27.8	27.8
Na_2EDTA	37.3	20.0	37.3	37.3
Organics				
Nicotinic acid	0.5	5.0	1.0	5
Pyridoxin-HCl	0.5	0.5	1.0	0.5
Thiamine-HCl	0.1	5.0	10.0	0.5
D-Biotin				0.05
Folic acid				0.5
myo-Inositol	100	1 000	100	100
Glycine	2.0			2
Sucrose	30 000	30 000	20 000	20 000

acetic acid) to ensure iron is available at high media pH (functional at pH values of up to pH 8.0).

4.1.4 Vitamins

The list includes thiamine-HCl, nicotinic acid, pyridoxine-HCl, *myo*-inositol, pantothenate, biotin, para-benzoic acid, cholate and choline chloride. These vitamins all generally function as important co-enzymes in reactions crucial to primary plant metabolism. Of all this group thiamine is essential to plant culture growth, while the remainder improve growth particularly at low cell concentrations.

Generally vitamins are included in the media in mM quantitites (*Table 4.1*).

4.1.5 Carbon source

Undifferentiated callus are normally deficient in chlorophyll and are often maintained in the dark or under low light intensities. They are therefore dependent upon an external carbon source. This is normally sucrose (supplied at 2–3% w/v) but it can be replaced with glucose. Other sources of C are not as effective as sucrose (e.g. fructose, lactose, maltose and starch).

4.1.6 Organic nitrogen

There is some benefit from including small amounts of organic nitrogen in the medium. This may be provided by casein hydrolysate (0.02–0.1%) or vitamin-free casemino acid or as a single amino acid such as glycine. Generally the single amino acids are inhibitory to growth so care must be taken if they are being added to the media to stimulate callus formation. The amino acids that are commonly included are glycine (2 mg ml l^{-1}), l-glutamine, asparagine, tyrosine (100 mg ml^{-1}), and l-arginine or cysteine (10 mg l^{-1}).

4.1.7 Plant growth regulators

Plant growth regulator levels are usually the most critical factor for successful callus or cell suspension growth and differentiation. The optimum concentrations of auxins and cytokinins differ from species to species. Auxins and cytokinins stimulate cell division and control cell differentiation and morphogenesis. Although naturally occurring auxins and cytokinins are available, for example indole-3-acetic acid (IAA), zeatin (Zea) and *N*-isopentenylamino purine (2iP), the most commonly used compounds are the synthetic auxins, 2,4-dichlorphenoxyacetic acid (2,4-D), indole-3-butyric acid (IBA), 1-Naphthalene acetic acid (NAA) and *p*-chlorophenoxyacetic acid (pCPA) and synthetic cytokinins, kinetin (K), 6-benzylaminopurine (BAP)). Other types of growth regulators, for example, the gibberellins and abscisic acid are not routinely used. *Table 4.2* lists common growth regulators used, typical concentrations and directions on how to prepare stock solutions.

The complete Murashige and Skoog (MS), Schenk and Hildebrandt (SH) [2] and Gamborg B5 (GB5) [3] media can be obtained from

Table 4.2. Plant growth regulators used in plant tissue culture media. Normal concentration range is 10^{-7}–10^{-5} M

Class	Name	Abbreviation	Mol. wt.	Preparation of stock solution	Comments
Auxin	p-Chlorophenoxyacetic acid	pCPA	186.6	All auxins dissolved in dilute NaOH or aqueous ethanol	IAA not normally as sole auxin source
	2,4-Dichlorophenoxyacetic acid	2,4-D	221.0		
	Indole 3-acetic acid	IAA	175.2		
	Indole-3-butyric acid	IBA	203.2		
	1-Naphthaleneacetic acid	NAA	186.2		
	2-Napthoxyacetic acid	NOA	202.2		
Cytokinin	6-Benzylaminopurine	BAP	225.2	All cytokinins dissolved in dilute NaOH or aqueous ethanol	
	N-Isopentenylaminopurine	2iP	203.3		
	6-Furfurylaminopurine (kinetin)	K	215.2		
	Zeatin	Zea	219.2		Zeatin not autoclave stable
Gibberellin	Gibberellic acid	GA$_3$	346.4	Dissolved in water	
Abscisic acid	Abscisic acid	ABA	264	Dissolved in aqueous ethanol	GA not autoclave stable

(see Appendix A), or each can be obtained as separate components. Reference to 'MS media' in the text refers to the use of MS inorganics (macro- and micro-inorganics), vitamins, organic nitrogen and sucrose. If only a portion of a media is used, this is always stated, for example, MS inorganics.

4.1.8 Agar

Where tissue cultures are maintained on solid medium, the medium is solidified with 6–10 g l^{-1} agar. The agar quoted is usually of high quality but this seems not to be critical for maintenance and differentiation of callus cultures. The exception to this is where haploid callus or protoplast-derived callus is being cultured when a high quality agar is required.

4.2 Additions to basic media

There are now numerous types of media, which vary slightly in their composition. *Table 4.1*, which includes four of the commonest media in use (MS [1], SH [2], GB5 [3] and Nitsch and Nitsch [4]) shows that all contain the same major groups of components. The media generally take the name(s) of those who devised them.

The MS media and SH are amongst the most commonly used in plant cell culture work. Both are regarded as 'high salt' media in view of their macro element composition as opposed to a 'low salt' media such as that of White [5].

Whenever possible, a fully defined media such as those in *Table 4.1* should be used. However, where a tissue is particularly recalcitrant, callus growth may be initiated or improved by the addition of complex additives. The composition of these compounds is variable and the source of active compound(s) not always known. Usually they are replaced as soon as possible by synthetic compounds. The additives commonly used are yeast extract (0.1–1.0 g l^{-1}), casein hydrolysate (0.1–1.0 g l^{-1}) and coconut milk (1–10%). The compounds provide a source of reduced nitrogen (casein hydrolysate) vitamins (yeast extract) and growth regulators (coconut milk). The yeast extract and casein hydrolysate can be bought directly from suppliers whereas the coconut milk must first be harvested from the ripe nuts. The decanted milk is checked for contamination by smelling it, any debris is filtered off, and the milk is then transferred to 500 ml vessels, autoclaved and stored in the deep freeze. When required, the milk is dissolved and refiltered.

4.3 Preparation and sterilization of media

Preparation of the nutrient media must always be carried out with care as mistakes can have very long-term consequences. The risks have been minimized since all of the standard media can be obtained in powdered form and at various levels of complexity, e.g. MS as the full media containing both the macro and micro elements, organic nitrogen and vitamins, or as MS basal salt medium composed of both the inorganic macro elements and organic micro elements. Thus it is possible to obtain separately, the inorganic macro elements, micro elements, organic nitrogen, vitamin supplements or any combination. Selected growth regulators may also be bought as prepared stock solutions. As it is sometimes necessary to manipulate individual components of the media, stock solutions may need to be prepared in the laboratory.

For ease of preparation, the media is divided into the following stock solutions: macro elements, micro elements, iron source, organic supplements, auxin and cytokinin. Each group is prepared as a separate stock solution according to the weights given in *Table 4.1*. In order to save time it is helpful to subdivide the stocks into volumes equivalent to a liter of medium, then store them frozen as separate small volumes. Because a large volume of stock solution is not being opened regularly there is less chance of contamination of the stock solutions.

The growth regulators (auxins and cytokinins) can present special problems and may need to be dissolved in a small volume of dilute NaOH, or ethanol, before making up to the final volume of the stock solution (*Table 4.2*).

One liter of complete nutrient media is prepared from the stock solutions, or commercial powder as follows:

1. The volumes of stock solutions (macro elements, micro elements, organic additives and growth regulators) are added to a 1 l conical flask on a magnetic stirrer, and 800 ml water added. Alternatively the weight of stock powder equivalent to 1 l of media is added to the flask and 800 ml of water added.

2. Sucrose is added to the solution which is stirred until all solids are dissolved, then the volume is made up to approximately 950 ml.

3. The pH will probably need a small adjustment with 0.5 M NaOH to give a final pH of 5.6.

4. The volume is made up to 1 l by transferring the volume to a 1 l measuring cylinder. (There is no advantage in using a 1 l volumetric flask to obtain an accurate volume measurement.) The solution is then transferred back to the original flask for complete mixing.

5. The medium is then transferred to conical flasks for autoclaving, that is 20 ml to 100 ml flasks and 40–50 ml to 250 ml flasks. The flasks are capped with aluminum foil then autoclaved.

6. Where solid agar medium is being prepared, the agar may be added to the medium before autoclaving. Agar is added to the medium in a saucepan then slowly warmed over a gas ring, the mixture stirred and brought to the boil briefly. The hot medium is distributed to the containers using the same volume of medium to container as liquid medium.

7. Alternatively, the liquid medium is distributed to 500 ml medical flats, the required weight of agar added then autoclaved. This mixes the agar and medium adequately and the liquid agar medium can then be distributed aseptically to the sterile containers if the temperature is above 40°C. Where the medium has cooled and set it is necessary to dissolve it again in a steamer before distributing it aseptically.

It is preferable to leave the media at 25°C for four days to check for bacterial infection before inoculation. If the medium is to be kept for longer periods, then it is better stored at 4°C.

References

1. **Murashige, T. and Skoog, F.** (1962) *Physiol. Plant.* **15,** 473–497.
2. **Schenk, R.U. and Hildebrandt, A.C.** (1972) *Can. J. Bot.* **50,** 199–204.
3. **Gamborg, O.L., Miller, R.A. and Ojima, K.** (1968) *Exp. Cell Res.* **50,** 151–158.
4. **Nitsch, J.T. and Nitsch, C.** (1969) *Science,* **163,** 85–87.
5. **White, P.R.** (1934) *Plant Physiol.* **9,** 585–600.

5 Initiation of callus cultures

In this chapter details are provided for the initiation of plant cell cultures. The stages in this procedure are the choice of explant, the method of surface sterilization and the composition of the medium (*Figure 5.1*).

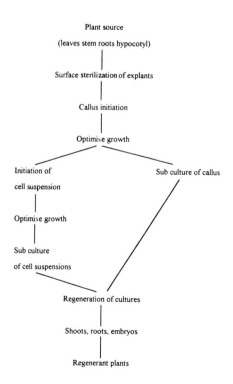

Plant source

(leaves stem roots hypocotyl)

Surface sterilization of explants

Callus initiation

Optimise growth

Initiation of cell suspension

Sub culture of callus

Optimise growth

Sub culture of cell suspensions

Regeneration of cultures

Shoots, roots, embryos

Regenerant plants

Figure 5.1. Sequence of stages in the initiation of callus and cell suspension cultures.

5.1 Source of explants for plant cultures

Callus can be obtained from explants of almost any part of the original plant, both from plant organs (e.g. roots, leaves and stems) and from specific tissue types or cells (e.g. pollen, endosperm, mesophyll). The research aims of the project may dictate which part of the plant is to be used as the explant. The important points to bear in mind when deciding upon the choice of explant are that the explant must contain living cells, that younger (juvenile) tissue contains a higher proportion of actively dividing cells and is therefore more responsive to a callus initiation program, that the parent plant must be healthy and free of any obvious sign of decay or disease, and that the plants from which the explants are obtained must be actively growing and not be about to enter a period of dormancy. It is important to bear in mind that the physiological status of the plant does have an influence on the response to the tissue culture program.

5.2 Sterilization methods to remove surface contaminants

Having identified a potential source of explants, the next step is to ensure that it is aseptic prior to it being placed on a callus initiation medium. Sterilization procedures are capable of killing all living cells, the sterilant should be in contact with the explant long enough to kill microbiological flora on the surface of the explant only, and not to destroy too many cells of the plant explant itself. This means that there is always an element of compromise in the choice of sterilant and the length of time of sterilization.

At the start of any tissue culture project it is always worth consulting and testing the published methods of sterilization. These methods may require modification in order to find one that works for a species or genotype that has not been cultured previously.

5.2.1 Seeds

To fulfill the above criteria, many research workers choose newly germinated seedlings as their choice of explant material since seeds are relatively easy to sterilize and can be germinated and grown under aseptic conditions. This choice of explant produces young vigorously growing tissue which is usually free of infection after sterilization of the seed coat and which contains a high proportion of actively dividing cells.

The successful method will depend upon the type of seed that is being surface sterilized. Seeds with rough surfaces usually require longer sterilization times and frequently need to be pre-treated with either solvents or to have detergents added to the sterilant to facilitate the wetting of the seed coat surface, thus ensuring that all potential microbiological contaminants come into contact with the sterilant. The following two protocols are a description of the methods used for medium-sized seeds (onion, soybean) and small seeds (tobacco, celery, alfalfa).

1. Medium-sized seeds are placed in a sterile conical flask containing a freshly made solution of a 10% (v/v) commercial bleach solution (e.g. Domestos). It is more convenient to contain the seeds in a tea strainer, or gauze bag. Concentrated solutions of sodium hypochlorite can be obtained from most of the major chemical companies but this concentrated solution will require a 1 in 10 dilution (Domestos or sodium hypochlorite does not require autoclaving prior to use).

2. Small seeds are briefly submerged (1–10 s) in absolute ethanol prior to the addition of a 10% (v/v) commercial bleach solution containing one or two drop(s) of Tween 80 (a detergent and surface wetting agent). Small seeds should be contained in a bag or tea strainer.

3. The seeds are completely covered by the sterilizing solution and the solution is swirled at regular intervals during the sterilization period of 15–20 min.

4. The hypochlorite solution is decanted and the seeds rinsed in three changes of sterile distilled water each lasting 5 min, after which time the seeds should be free from any smell of hypochlorite and are ready to be plated on to media.

5. After the hypochlorite solution has been added, all subsequent manipulations and treatments have to be conducted aseptically if the seeds are to remain free of contamination prior to their culture.

6. The seeds are poured onto the surface of a sterile petri dish, or the bag opened on the surface of a sterile petri dish, then the seeds

transferred individually to a Universal vial containing 5–8 ml of germination medium (0.1 MS medium, 1% sucrose, and all growth regulators omitted).

7.   The surface sterile seed may be placed directly onto a callus-inducing medium so that germination and callus induction occurs in one stage.

If the standard sterilization methods are not successful for the seed of any particular species or genotype, then the duration of exposure to the sterilant and concentration of sterilant must be altered using a range of times and concentrations.

5.2.2 *Vegetative tissues*

Explants can be prepared from leaves or stems from plants that have been grown in the greenhouse or in the field. The main problem with such material source is the increased risk of endogenous microbiological contaminants that the surface sterilization methods can not eradicate. Source material for the explants must be healthy and free of any obvious sign of infection or insect damage.

1.   The plant material is washed to remove any large particulate matter such as soil, then dipped in 70–100% ethanol to facilitate wetting of the hydrophobic plant surfaces.

2.   The plant material is surface sterilized in sodium hypochlorite solution (5–20%) for a suitable period of time (5–10 min) then washed in at least three changes of sterile distilled water.

3.   If the explant is a section of stem or petiole, any cut surfaces that have been in contact with the sterilant should be removed before transferring to sterile medium. This is necessary because the sterilant will kill exposed plant cells which may then form a barrier to the uptake of the nutrient medium.

4.   The explant is sectioned and transferred to callus-inducing medium. Normally, 10 mm stem sections or 10 mm diameter leaf portions are used.

Where the plant material is storage organs such as tubers or swollen roots which have been in prolonged contact with the soil, sterilization may be a problem. The plant material is washed and surface-sterilized as described above. The skin is removed under aseptic conditions and the internal tissue is excised with a sterile cork borer (5 mm diameter) and 5 mm long sections plated on to a suitable medium. If the plant organ is particularly large, it can be sub-divided into smaller sections prior to sterilization as long as the surface layers of cells that have been in direct contact with the sterilant are not plated out.

5.3 Initiation of callus from explants

Seedlings, germinated from surface sterile seeds, provide a source of roots, hypocotyl, stem and leaves and non-seed sources provide leaves, stems and storage tissue. These are sectioned as described below and placed on variations of the media described in Chapter 4 (*Figure 5.1*).

1. Seedlings are removed from the containers and placed on the surface of a sterile petri dish. The different parts are then excised into uniform sizes and portions are placed individually onto an agar medium in vials or petri dishes. The hypocotyl or stem is the most uniform part of the seedling and is the most appropriate part as 10 mm sections to use in any experimental program.

2. Uniform explants are excised from surface sterile vegetative structures then pressed into the surface of agar medium in petri dishes. This action ensures that the explant (e.g. leaf disks), particularly the cut edges, are in good contact with the medium.

3. The petri dishes are sealed with Parafilm which prevents desiccation of the medium and also entry of culture mites which introduce contamination.

4. Tissues are incubated at 25°C in the light, or dark, depending on the requirements of the genotype.

Assuming a suitable medium is used, most callus will be initiated from the cut surfaces within 3–8 weeks. Initiation of tomato callus normally takes 2–3 weeks but initiation of onion callus can take as long as 7–8 weeks. It is important that the explants are checked regularly and any contaminated cultures removed to prevent the spread of infection. If after 8 weeks no callus is visible on the explants, then it can be assumed that the attempt to initiate callus has been unsuccessful. The sterilization method and the composition of the medium should then be re-examined.

5.3.1 Screening of media to establish optimum initiation of tissue cultures

Before attempting to initiate callus cultures, the literature must be consulted to establish whether there has been a successful callus initiation previously for the species and genotype under consideration. There are only a limited number of genotypes of a crop species that have been cultured successfully. If tissue cultures are required from a species or genotype not previously in culture, then it may be necessary to modify a medium that was used for a related species or genotype.

The standard approach would be to start with one of the defined media, for example, MS, SH or GB5 (Chapter 4) and experiment by manipulating both the concentration of auxin and cytokinin and the ratio between the two groups of plant growth regulators. The practise is to identify a range of six concentrations of auxin (e.g. 0, 0.1, 0.25, 0.5, 1.0, and 5.0 mg l⁻¹) and six concentrations of cytokinin (e.g. 0, 0.1, 0.25, 0.5, 1.0, and 5.0 mg l⁻¹) and test them in all thirty six combinations in a Latin Square arrangement to see which combination produces the optimum development of callus (*Table 5.1*). The range of concentrations will be determined by the concentration of auxin and cytokinin in media used to initiate callus of a related species or genotype. If the response to these combinations is poor, then other growth regulator combinations will have to be investigated using the same approach.

Each media combination is replicated 10 times since the response is normally quite variable. The response is assessed on a semi-quantitative basis by giving the amount of callus production on each explant a score of 1–4. The combination of auxin and cytokinin with the largest score is the one that is most suitable for callus initiation.

Normally, modifications are only made to the growth regulator composition rather than the other major components or individual

Table 5.1. Latin Square arrangement for testing media combinations of auxin and cytokinin

Cytokinin conc. mg l⁻¹	Auxin conc. mg l⁻¹					
	0.0	0.1	0.25	0.5	1.0	5.0
0.0						
0.1						
0.25						
0.5						
1.0						
5.0						

elements. However, in some cultures such as haploid or protoplasts cultures there may be an alteration to other parts of the medium (see Chapters 8 and 9).

5.3.2 Effect of media components on initiation of tissue cultures

The Latin Square analysis above will show what effect growth regulator components of the media will have on the tissue culture response. If all components of the media are varied in turn, their relative importance can be assessed in the same way. For example African Violet provides a good model system for this type of analysis. Young leaves of African Violet are surface-sterilized by a brief immersion in 70% ethanol, 1 cm squares cut from each leaf are immersed in 10% Domestos solution for 5 min and then washed in sterilized distilled water 3–4 times. Three squares are pressed onto medium contained in 9 cm petri dishes. These are sealed with Parafilm then left in a tissue culture room in the light at 25 °C for six weeks. The medium composition has a control (MS 0.5 mg l^{-1} NAA, 0.5 mg l^{-1} BAP, 3% sucrose and 10 g l^{-1} agar) and treatments in which in turn the sucrose, the cytokinin (BAP), auxin (NAA) and MS salts are varied by omission, or included at a reduced, or enhanced concentration (*Table 5.2*).

Table 5.2. Modifications to media for culture of African Violet

Treatments		MS salts	NAA	BAP	Sucrose
Control		4.71 g l^{-1}	0.5 mg l^{-1}	0.5 mg l^{-1}	3%
Vary Sucrose					
2		"	"	"	0
3		"	"	"	1%
4		"	"	"	5%
Vary BAP					
5	0	"	"	0	3%
6	1/5th	"	"	0.1 mg l^{-1}	3%
7	x 10	"	"	5.0 mg l^{-1}	3%
Vary NAA					
8	0	"	0	0.5 mg l^{-1}	3%
9	1/5th	"	0.1 mg l^{-1}	"	3%
10	x 10	"	5.0 mg l^{-1}	"	3%
Vary MS Salts					
11	0	0	0.5 mg l^{-1}	"	3%
12	1/10th	0.47 g l^{-1}	"	"	3%
13	x2	9.52 g l^{-1}	"	"	3%

The results will show that in the control there is multiple adventitious bud formation leading to a large number of small shoots on the upper surface of the leaf not in contact with the medium. Absence of sucrose inhibits this production and shoot production is limited at low sucrose and comparable with the control at high sucrose. At zero and low BAP, callus forms at the leaf surface in contact with the medium but at high BAP there are a large number of shoots. At zero and low concentrations of auxins there is a stimulus to shoot formation and at high concentrations large amounts of roots are formed. At low and zero MS salts there is no growth at all. These variations in the medium very clearly demonstrate the importance of a carbon source and an inorganic salt source and the effect of the auxin cytokinin balance on the level of callus, or root and shoot production.

5.4 Presence of endogenous contaminants

Endogenous contamination arises from microbial infections present within the internal tissue of the explant and consequently cannot be eradicated by the usual surface sterilization procedures. Endogenous infections are more common in cultures that have been initiated from explants derived from material grown in the greenhouse or in the field. When using this plant material as the source of explants it is important to maintain the plants in a healthy condition, ensure regular feeding to maintain vigorous growth, to treat with fungicides and insecticides if applicable and when selecting material for explants it must be free of infection. The use of seedlings germinated under aseptic conditions helps to reduce this problem as long as the seed itself does not have an endogenous infection. It is also important to use a good aseptic technique.

If these precautions do not eliminate the problem the final resort is to add antimicrobial agents such as antibiotics (Claforan or Geopen) to the media. A range of concentrations will need to be tested but a starting point could be 200–400 mg l^{-1}. These compounds reduce the growth vigor of the cultures but their effect is only temporary and the culture will return to a more vigorous growth once the antibiotics are removed.

6 Growth of callus and cell suspension cultures

Having initiated callus on explant tissue, the next stage of the process is to isolate the callus and maintain it as a rapidly growing tissue mass. It can then provide a source of uniform cells for a variety of purposes, such as for regeneration, isolation of protoplasts, or *in vitro* screening for resistant mutants. If a source of uniform callus is required, then it essential that the callus can be subcultured with no loss of vigor over a period of time. Examination of callus formation on the initial explants will show that the response to the initiation medium is quite variable. Not all of the explants will have produced callus, and where callus has been initiated, the color, friability and amount will vary. The reason for this wide variation in response from apparently uniform explants is not clear.

6.1 Maintenance of callus cultures

Subculture of callus should always take place in a laminar flow cabinet or sterile room. A short spoon spatula to sub-divide and transfer the callus is kept in a wide-based glass container half-full of ethanol. Sterile transfer procedures are then as described in Chapter 3. If Universal vials are being used then the spatula may need to be filed down so that it can be inserted easily into the vials and the front edge sharpened to cut any callus that is not friable.

At the earliest stage of callus initiation, only small portions of friable callus may be available for transfer to the new medium. In a transfer of stock cultures of callus, portions approximately the size of a small pea are normally transferred to a choice of containers which may be either petri dishes, Universal vials with screw tops, 50–100 ml Erlenmyer conical flasks capped with aluminum foil, 9 cm petri dishes or plastic disposable containers.

Callus cultures are maintained in a growth room under low intensity fluorescent lighting with a diurnal cycle of 12 h in the light or dark. The dark or light requirement would need to be established for each species and genotype. Cultures are maintained at a temperature of 25 °C for most species. Growth is slowed down at a lower temperature and may cause browning in a tropical crop species such as cocoa. Callus cultures are subcultured under these conditions every 2–6 weeks depending on the growth rate.

6.1.1 Improvement in callus growth

If callus initiation on the explant is vigorous and has produced portions of friable lightly colored callus 2–5 mm in diameter, these can be subcultured by excising and transfer of the isolated callus to fresh medium under the same conditions as before. Alternatively if the callus is < 2 mm diameter, the callus plus explant is better transferred to fresh medium for a further period of growth until a separate portion of callus can be isolated. Normally callus is transferred to a maintenance medium which has the same composition as the callus initiation medium. If the maintenance medium is unsuitable, the callus will become brown and exhibit slow growth. Sometimes the delayed appearance of contamination can contribute to this poor growth.

If poor growth is not a result of contamination then two approaches can be taken to overcome the problem; either the more vigorously growing and friable callus pieces are selected and subcultured routinely, or the medium is changed.

1. Before each subculture only friable and light-colored callus is isolated from the callus. These portions, however small, are then used as the new inoculum. This selection of light-colored vigorously growing callus is normally practised during the long term maintenance of all callus cultures.

2. If a more rapid improvement is required then the concentration of the growth regulators (auxin and cytokinin) of the initiation medium must be modified. The concentration of the auxin and cytokinin are varied above and below the concentration used in the initiation medium in a Latin Square arrangement (Chapter 5). The growth of uniformly sized replicate callus (15–20 mg, approximately pea-sized) is then assessed on a semi-quantitative basis after two successive subcultures (each 3–4 weeks long) on these combinations. A numerical assessment is made based on a gradation in size (1–4) with 4 being the most rapid growth, and

the intensity of brown color (1–4) with 4 being the lightest color. The treatment with the highest score is then the preferred combination.

6.1.2 *Growth measurement of callus*

Growth of callus is measured by assessing the fresh weight and dry weight of the cultures. Fresh weight is measured by transferring the callus to a weighed aluminum square then the combined weight is measured as rapidly as possible after the callus is removed from the tissue culture containers, because the callus loses water, and therefore weight, very quickly. Once the fresh weight is measured, the aluminum square is closed carefully over the callus without sealing it, then the foil envelopes are kept in an oven at 80°C for 2 days. Both foil and callus are reweighed and the weight of callus is obtained by subtraction. Because of the rapid and uncontrolled weight loss of the fresh callus, the dry weight is the more reliable criterion.

A typical growth curve for fresh weight and dry weight of cocoa callus is shown in *Figure 6.1*. The pattern of growth is similar for all species. There is an initial lag phase followed by a period of rapid growth which may approach exponential, then a period of declining growth

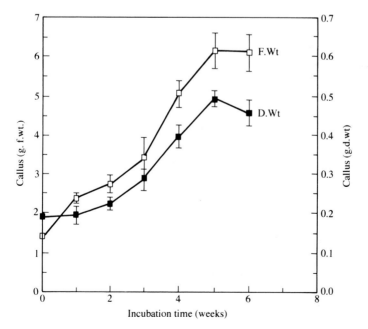

Figure 6.1. Fresh and dry weight of cocoa callus during a subculture period in 20 ml agar medium in 100 ml conical flasks.

followed by a stationary phase and eventually a decline in biomass as the cells in the culture begin to senesce.

Some callus cultures have been maintained for many years under these conditions but the culture is not necessarily stable. Small changes in culture conditions such as the length of the subculture cycle, or temperature can affect the culture by imposing a different set of environmental conditions. Because the cells in the culture are not uniform, the new set of conditions may select cells from this population which are more adapted to the changed conditions. If uniform cultures are required it is important that the conditions are kept as constant as possible. Even though the conditions are stable, long-term cultures may accumulate variation as a result of non-lethal somatic mutations and changes in ploidy. These changes as somaclonal variation have been exploited to improve disease resistance and environmental resistance of a number of crop species (Chapter 11).

6.1.3 Contamination in callus cultures

Contamination is a major problem and if undetected can lead to the loss of experiments and stock cultures. It is important to inspect carefully each container for signs of contamination on the medium and the callus before any subculture is attempted. Bacterial contamination will appear within a week of subculture as a faint white ring on the agar around the callus isolate. Any fungal contamination will be seen as hyphae on the callus and the surrounding agar two weeks after the subculture. Yeast contamination is more difficult to detect without examining the callus under the microscope but it usually gives a white appearance to the agar surrounding the callus.

In all instances of contamination there is usually a reduction in growth of the callus and a browning of the tissue. The problem is usually avoided by early detection and elimination of any suspect flasks. If an endogenous contaminant is suspected then treatment of the cultures with antibiotics is a possibility (Chapter 5).

6.2 Initiation of cell suspension cultures

Callus does provide more accessible uniform cells than the intact plant, but the tissue is not uniform since only the base of the callus is exposed to the medium and the callus mass may contain cells at all stages of development. The alternative is to use cell suspensions. These can be grown in bulk, show a faster growth rate, and all cells are exposed uniformly to the medium. Cell suspensions are preferred as a source of protoplasts for fusion or genetic manipulation, for large scale embryogenesis and for commercial production of secondary products. Cell suspensions are normally initiated by transferring callus to liquid media of the same composition as the callus medium and gently agitating the suspension on a horizontal shaker at 100 rpm and 25 °C. The callus will often break up and form a suspension of cells which consists of cell aggregates of varying size and cell number, and also single cells (*Figure 6.2*).

There are often problems in initiating a cell suspension as the original callus lumps may not break up, or may turn brown very quickly. This problem can be resolved by taking the same approach as that taken with recalcitrant callus. The procedure is to either select those cells that will grow in the liquid medium or to modify the composition of the medium.

1. In the approach based on selection, 1 g portions (2–3 pea-sized lumps) of the most friable and pale-colored callus are placed in a relatively small volume of liquid medium (10 ml in a 50 ml conical flask) of the same composition as that used for the maintenance of callus. Cells in a larger volume of medium suffer a loss in viability as a result of leaching. The flask is placed on a horizontal shaker and rotated at 100 rpm under standard growth conditions. After 1–2 weeks, a further small volume of fresh medium (5–10 ml) is added to the existing flask. This addition of fresh medium is continued until it is apparent from the soup-like appearance of

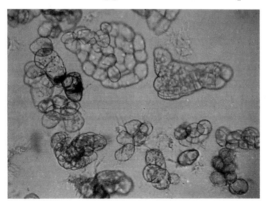

Figure 6.2. Suspension of celery cells showing single cells, small cell groups and cell aggregates.

the suspension that the cells are breaking off the callus and dividing in the medium. It may be necessary to transfer the cell suspension to a 100 ml conical flask. The suspension can then be subcultured into fresh medium but the volume of inoculum to medium is again kept to 1:1. This procedure is continued until the growth of the cell suspension shows no evidence of a decline, but continues to form a thick suspension.

2. The alternative is to modify the medium by changing the concentration of the auxin and cytokinins on the basis of a Latin Square design (see *Table 5.1*). Approximately four pea-sized portions are added to a 100 ml conical flask containing 20 ml medium, then the flasks are placed on a horizontal shaker for 3–4 weeks. The growth of the cell suspension is estimated on a semi-quantitative basis using density (1–4) with 4 being the most dense, and color (1–4) with 4 as the lightest color. After assessing the cell suspensions at the end of the first subculture, the medium is drained from each flask then a volume of fresh medium of the same composition is added to each flask containing the residual cells. The cell suspensions are then assessed as before. Once a medium has been identified that will support the sustained growth of the suspension, the inoculum to fresh medium ratio can be reduced to 5 ml cell suspension to 20 ml fresh medium in a 100 ml conical flask. Where larger amounts of biomass are required, an inoculum of 10 ml is added to 50 ml medium in a 250 ml conical flask.

6.2.1 Maintenance of cell suspensions

All subculture of cell suspensions should take place in a sterile room or laminar flow cabinet. The arrangement of containers within the cabinet is the same as for callus subculture and the same aseptic techniques should be observed (Chapter 3). The cell suspension is subcultured by using a wide-bore glass or plastic pipette. This is essential as the cells settle out very quickly and will block the outlet of a narrow-bore pipette. The flask due for subculture is swirled to bring the cells into suspension then an aliquot is withdrawn into the pipette using a bulb, and the aliquot is released into the new container which is then capped. The same pipette can be used to withdraw aliquots from the stock flask but a new pipette must be used when the suspension from a new stock solution is used. This avoids any possibility of transferring infection to all the new containers.

6.2.2 Growth measurement of cell suspensions

Growth of cell suspensions is measured by filtering the cell suspension through a single layer of Whatman No. 1 filter paper lining the inside of a Buchner funnel with the funnel attached to a vacuum line. To avoid flooding the surface of the filter paper, a measured volume of cell suspension is dropped slowly from a wide-bore pipette onto the surface of the paper so as to allow the medium to be removed in a controled manner. The vacuum is maintained until the excess medium is removed as judged by a lightening of the colour in the cell mass. The cells are then scraped carefully off the surface of the paper, avoiding removing any of the paper, and transferred onto a weighed aluminum foil square and the combined weight is measured. The foil is folded over the tissue culture as before and placed in an oven at 80 °C for two days and then the combined weight is measured.

An additional method of measuring the growth of cell suspension is by packed cell volume (PCV). A measured volume of the suspension is transferred to a graduated test tube which is centrifuged at low speed (100 rpm for 10 min) and the volume of the cells expressed as a percentage of the volume of cell suspension used. A vigorously growing cell suspension will show a 40–50% PVC. In the past, cell numbers have been used as a measure of growth of cell suspensions, but it is often not possible to separate the cells within the cell aggregates of the cell suspensions adequately, making the technique difficult to use. The most satisfactory method of measurement is the fresh weight and dry weight of the cell suspension.

The growth of the cell suspension shows a similar growth curve to the callus with an initial lag phase, an exponential phase and a stationary phase (*Figure 6.3*) The major difference is that the growth cycle is completed more rapidly (1–3 weeks) and the phases are more obvious than in the callus. Culture senescence sets in rapidly after the stationary phase so that the cell suspension must be subcultured within the 3 week period, otherwise the culture will be lost. Provided that the same conditions of subculture and culture conditions are maintained routinely, the cell suspension will show a constant growth rate. Stock cultures are maintained under these constant conditions to provide uniform cells. The growth curve can be altered however by altering any of the culture conditions. A lower inoculum density will extend the lag phase and a higher density will shorten it. An inoculum of exponentially dividing cells will also reduce the lag phase. A higher shaker speed will increase the growth rate, but if the speed is too low, the cells will die. A higher sucrose concentration will extend the growth period and a lower concentration will reduce it.

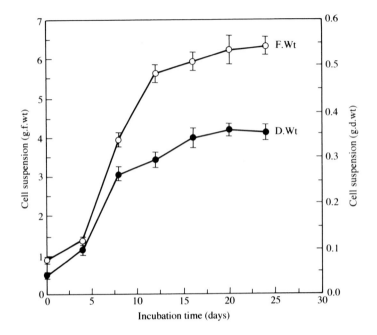

Figure 6.3. Fresh and dry weight of cell suspensions of cocoa in 20 ml medium in 100 ml conical flasks.

The cell suspension even under constant conditions of culture still consists of a variable population of cells. This source of variation provides the basis for *in vitro* cell selection (Chapter 11). After a period of prolonged subculture, cultures do change. This can be seen most obviously as an increase in the number of cells in the culture showing aneuploidy and polyploidy. There are also other less obvious changes to the genome which contribute to somaclonal variation.

6.2.3 Contamination in cell suspensions

One of the major problems in maintaining cell suspensions is the incidence of contamination which, if left undetected, may be transferred to all the cultures. The contamination may be from aerial contaminants if the transfer cabinet is not effective, or the screen is due for a change, or poor sterilization of the flasks and pipettes. It is essential to check all cultures before use for contamination. With practise even low levels of contamination can be detected.

Bacterial contamination gives a blue haze to the medium. This blueness can also be caused by the release of wall polysaccharides into the medium. Bacterial contamination is distinguished from this by examining the meniscus the medium makes with the glass flask. If this is clear then the culture is not contaminated. However, if the meniscus still appears blue then the culture contains a bacterial contaminant. All suspect cultures must be rejected otherwise contamination will spread rapidly through both experiments and stock cultures. Fungal contamination will often appear as balls of mycelium in the medium and contamination by yeast as a creamy color to the cell suspension. Microscopic examination of the liquid culture will confirm the presence of contamination. Even in low levels of contamination, the plant cells will appear brown and the culture will show a reduced growth.

7 Regeneration of tissue cultures

When plant tissue is placed on a nutrient medium containing a high concentration of growth regulators (auxin and cytokinin), the cells become de-differentiated and callus is formed. Conversely, callus tissue on a medium in which the concentration of the growth regulators is reduced, or the composition altered, or the growth regulators are omitted altogether, will often form large numbers of re-differentiated structures. These may be embryos (embryogenic path), shoot buds (caulogenic path) or roots (rhizogenic path). Not all species regenerate readily from undifferentiated tissue cultures, however. The cereals (wheat, barley, maize and rice), the forage legumes (alfalfa and soybean) and many woody perennials have proved in the past to be particularly recalcitrant, but now many of these species can be regenerated from callus and even from protoplasts. Often the response of callus tissue to a regeneration medium is strongly genotype-dependent, so that a strategy that is suitable for one variety may not work for another. In these cases, a successful medium can only be used as a guide for a variety that has not previously been in culture.

The important stages in all the regeneration methods are the source of explant tissue, the regeneration medium, the maintenance of regeneration activity of the cultures and finally the method of transfer to soil and hardening of the regenerants for growth in the glasshouse and field (see *Figure 5.1*). These aspects will be examined in more detail in this chapter. Detailed protocols for regeneration of many species can be obtained from the literature and this source should be consulted.

7.1 Development of somatic embryos

In the intact plant, the embryo develops from a single-celled zygote by rapid cell division to form a multicellular mass, the globular stage and a suspensor which attaches the embryo to the parent plant. The embryo then shows a more obvious morphological differentiation by formation of a heart-shaped stage, then this shape becomes more polarized to form the torpedo stage in dicots and the cylindrical stage in monocots. The two short arms of the torpedo embryos are the future cotyledons which enclose the presumptive stem apex at their base and at the opposite end is the presumptive root apex. At this stage the zygotic embryos show signs of internal organization with vascular strands containing recognizable xylem and phloem running from base to apex then separating into the cotyledons.

The origin of the tissue culture derived somatic embryos on callus, or on cell aggregates in cell suspension, is less clear. Examination of the surface cells of embryogenic callus of celery for example suggests that potential meristematic cells are characterized by a much denser cytoplasm than the surrounding cells. In cereal callus, the embryogenic cells also have a smaller, more densely cytoplasmic structure. Single densely cytoplasmic cells are thought to be the starting point for the embryogenic or caulogenic process, so that somatic embryos and buds are likely to be of single cell origin. The single cell origin of somatic embryos is important since an *in vitro* selection program using regenerating tissue selects for embryos or buds of which each will consist of one genotype rather than a chimera.

Development of the somatic dicotyledon embryos is similar initially to that of the zygotic embryos (*Figure 7.1*). Somatic embryos show an undifferentiated globular form (*Figure 7.2a*) then a morphologicaly distinct heart-shaped stage (*Figure 7.2b*) and a polarized torpedo stage (*Figure 7.2c*). The development of somatic and zygotic embryos then diverge. The zygotic embryos show a massive expansion of the cotyledons in dicotyledons and expansion of the endosperm in monocotyledons, and finally the tissues dehydrate and become the resting structure, the seed. In contrast, torpedo dicot or cylindrical monocot somatic embryos develop directly into young plants with no intervening seed stage.

7.2 Induction of regeneration in tissue cultures

The medium required to initiate regeneration of the callus may contain the same auxin and cytokinin compounds as the callus induction medium but a reduced concentration may be required. The optimum combination of growth regulators for regeneration of callus

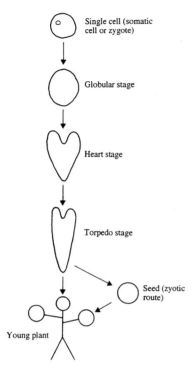

Single cell (somatic cell or zygote)

Globular stage

Heart stage

Torpedo stage

Seed (zyotic route)

Young plant

Figure 7.1. Stages in zygotic and somatic embryo development in dicotyledonous plants.

Figure 7.2a. Globular embryos from celery tissue cultures.

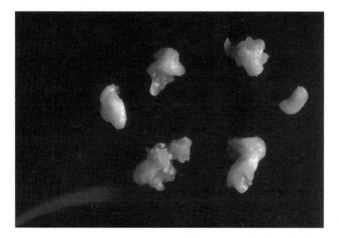

Figure 7.2b. Heart-shaped embryos from celery tissue cultures.

can be identified using a Latin Square arrangement of concentrations (Chapter 5), using the same auxin and cytokinin as in the callus maintenance medium for the genotype.

1. A convenient range of concentrations of auxin (0.0, 0.1, 0.5, 1.0, 5.0 and 10.0 mg l⁻¹) in combination with the same range of cytokinin concentrations is selected. Recently initiated callus is placed as 3–5 portions (10–15 mg per portion) on 20 ml medium in petri dishes (9 cm diameter), or as single portions on 8 ml agar medium

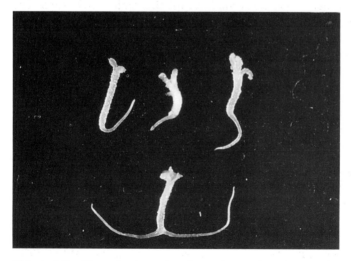

Figure 7.2c. Torpedo-shaped embryos from celery tissue cultures.

in Universal vials. Each of the 6 × 6 treatment combinations is normally replicated 10 times to take into account the wide variability in response of the initial callus explants. It is important that the callus is derived from meristematic explants such as young leaves, seedling tissue, mature and immature embryos and immature inflorescences and has not been subcultured for a long period.

2. If a caulogenic or an embryogenic regeneration response is identified in one or more treatment combinations, then the regenerating tissue is subcultured for a further period in this combination to confirm the response. Because the concentrations of auxin and cytokinin may be quite wide, the second subculture could include a more restricted range of concentrations in the range that was found to be effective.

3. In the absence of a regeneration response, it will be necessary to examine alternative combinations of auxin and cytokinin and also sources of the explant. Auxins that have been used in tissue regeneration are 2,4-dichlorophenoxyacetic acid (2,4-D), 2,4,5-trichlorophenoxyacetic acid, 1-naphthaleneacetic acid (NAA), indole-3-butyric acid (IBA) and indole-3-acetic acid (IAA) and the cytokinins *N*-isopentenylaminopurine (2iP), 6-benzyl-aminopurine (BAP), kinetin (K) and zeatin (Zea) (Chapter 4).

Large numbers of species of dicots and monocots, annuals and perennials have now been regenerated from tissue culture. Although there are wide differences in approach and many genotypes are still recalcitrant, progress has been made in all the major food crops. The carrot, which has been used extensively as a model plant tissue culture, shows clearly the general approach that needs to be adopted when initiating a regeneration program. Modifications to this basic approach are described for a number of other crops.

7.2.1 Carrot embryogenesis

In the early work on tissue culture regeneration in carrot, the explants were surface-sterile plugs of the carrot root. Because the initial response was slow, the preferred explant source now for the initiation and regeneration of carrot tissue cultures is to use portions of seedlings [1].

1. Seeds are surface-sterilized (Chapter 3) and are then transferred individually to Universal vials and germinated in the light at 25°C.

2. When the hypocotyl is 5 cm high, 1 cm sections of the stem, or root of the seedling are placed onto a nutrient medium (MS agar medium, 0.5 mg l^{-1} 2,4-D, 0.6 mg l^{-1} kinetin and 2% sucrose) to initiate callus [1].

3. On this high level of auxin, callus appears on the surface of the explant within 3 weeks and at 6 weeks the callus can be separated from the explant and be subcultured routinely on the same medium every 4 weeks. The response from each seed will not be identical and even at this stage it is possible to select the more rapidly growing cultures.

4. The callus is separated from the original explant then transferred to a medium in which the auxin level is reduced to 0.1 mg l^{-1}, (with all the other components remaining at the same concentration). Many of the cultures will show signs of differentiation after 3 weeks. A further subculture on the medium containing reduced auxin levels will enhance the number of regenerating structures.

5. The form of these structures can be examined most easily by transferring the regenerating culture into a petri dish containing some water then gently breaking up the callus. If the suspension is now examined under a binocular microscope, the undifferentiated callus will appear as white translucent masses. Amongst these cells there will be white globular and heart-shaped stages, early torpedo forms in which greening is just beginning. These differentiated structures can be identified both by their shape and their more dense white appearance compared with the translucent callus cells. There is also considerable variation in size within the three embryonic forms.

6. The regeneration sequence in the culture appears to be held at this early embryonic stage by the presence of low concentrations of auxin in the medium. It is possible to maintain the culture in this regeneration phase almost indefinitely by repeated subculture, thus providing a constant source of regenerating tissue cultures.

7. If the undifferentiated callus is transferred directly to a medium in which the auxin is omitted altogether, differentiation will proceed beyond these early stages. Within three to six weeks on the new medium the cultures show green spots and larger green structures. Examination of the cultures as before will reveal that the cultures contain large green torpedo-shaped embryos often

with elongated and misshapen cotyledonary leaves, as well as a small number of globular and heart-shaped stage embryos.

The pattern of development suggests that the high auxin levels (0.5 mg l^{-1}) required to initiate callus were inhibitory to any differentiation. Lower levels of auxin (0.1 mg l^{-1}) while initiating differentiation were still inhibitory to its complete expression. It was only in the absence of auxin and cytokinin that differentiation was allowed to proceed to completion. The torpedo embryos in the culture on the growth medium develop beyond this stage if the entire culture is subcultured onto fresh medium in a larger wide-topped container (Chapter 5). The additional space stimulates the embryos to grow into numerous plantlets with distinct small stems and leaves. These highly differentiated cultures can be maintained *in vitro* by dividing the culture into smaller portions and transferring them to an auxin-free medium. This procedure allows the individual plantlet to develop in a less competitive environment and to a stage when it is suitable for transfer to soil.

7.2.2 Tobacco shoot morphogenesis

Tobacco is commonly used as a tissue culture but it has a different regeneration sequence to carrot since it tends to form buds rather than embryos.

1. The initial part of the procedure is to produce large amounts of surface-sterile leaf tissue which can then be maintained with the minimum of effort. Tobacco seeds are surface-sterilized then single seeds are placed in 1 ml MS inorganic medium, 0.4 mg l^{-1} thiamine, 100 mg l^{-1} *myo*-inositol and 3% sucrose contained in each well of a 25 well Sterilin plate. The plates are sealed with Parafilm and incubated in the light at 25°C.

2. Germinated seedlings are excised above the hypocotyl and the stem tip placed vertically in 25 ml MS agar medium in a 250 ml Sterilin container and incubated for 6–8 weeks. These shoot tips extend to yield 1–3 side shoots each of which can be excised and subcultured as above.

3. Disks (1 cm) diameter are removed from expanding leaves (2.5–3.0 cm long) and incubated on a callus induction medium (MS medium, 1 mg^{-1}, 2,4-D, 1 mg l^{-1} kinetin and 3% sucrose) in petri dishes in the light. After 3 weeks, callus forms on the edge

of the disk and at 6 weeks can be transferred independently to the same medium and subcultured every four weeks.

4.   The callus can be regenerated by transferring portions to petri dishes containing a regeneration medium (MS medium, 0.5 mg l^{-1} IAA, 1.0 mg l^{-1} BAP and 3% sucrose) in which the composition of the growth regulators is altered. Regeneration is initiated on this medium via a caulogenic route since large numbers of buds appear on the surface of the callus.

5.   The buds enlarge into plantlets which are transferred individually to 250 ml Sterilin containers to continue their normal development. Finally the plantlets are transferred to soil.

7.2.3 Soybean embryogenesis and shoot formation

In other species, induction of regeneration may require a more complex approach to achieve the transition from explant to callus, embryos or buds and finally to plantlets. It is essential in many of these more recalcitrant species that the starting material is potentially embryogenic, such as immature zygotic embryos. The stage of the development of the zygotic embryo at which it is isolated and tissue cultures initiated is important to the success of the regeneration program and may vary with species and variety. The soybean is an example where the immature embryo is used as an explant. Investigations have shown that there is an optimal stage of development of the embryo when it should be excised for tissue culture initiation and regeneration. In the soybean, the stage of development of the immature embryo can be identified by the size of the immature seed. When the immature seed is approximately 4–6 mm long, the immature embryo provides an explant which gives a positive regeneration response for a wide range of varieties of soybean. The regeneration sequence consists of three separate stages involving three media [2].

1.   Immature embryos are excised aseptically from seed that is 4.0–6.0 mm diameter then transferred to petri dishes and maintained in the dark for 3 weeks. There is an initial induction phase where the medium is designed to stimulate either an organogenic path (MS medium, 43.0 μM NAA, 5.0 μM thiamine, 0.03 mM nicotinic acid and 3% sucrose) or an embryogenic path of regeneration (MS major inorganic salts, 4 × MS minor inorganic salts, Gamborg's B5 vitamins, 13.3 μM BAP, 0.2 μM NAA, 5.0 μM thiamine and 12 mM proline).

2. The cultures are then transferred to a common regeneration medium for three weeks (MS basal inorganic medium, 1.7 µM BAP and 0.2 µM IBA, 3% sucrose) in the light which allows the particular induced pathways to be expressed. The two types of development are easily distinguished by the fact that the embryos at a late torpedo stage each possess two cotyledonary leaves which are simple elongated structures whereas the buds are more complicated presumptive shoot meristems. Both embryos and buds appear to arise directly from the surface of the immature embryo and not from an obvious callus phase.

3. When the buds or embryos are approximately 0.5 cm long they are transferred to a petri dish containing an agar growth medium (MS basal medium, 3% sucrose) which is the same as the regeneration except that it contains no auxin or cytokinins. This medium stimulates normal development of both embryos and buds into plantlets. When 1.0 cm high the plantlets are transferred to growth medium in larger containers in order to stimulate leaf and root growth in preparation for their transfer to soil (R. Tareghyan, personal communication).

7.2.4 *Rice embryogenesis*

Cereals and grasses which provide the major source of food and forage species were difficult to regenerate from tissue culture, but once it was found that the most suitable source of explant tissue was the immature embryo (or less commonly young inflorescences and young leaf bases), somatic embryogenesis became more routine. There are also two types of callus, one potentially embryogenic and the other not. The approach to regeneration is similar to that in dicots since there is an induction stage with high auxin followed by a regeneration stage with reduced auxin and finally a growth stage with no auxin and reduced inorganics. The example of rice has been given [3] but that of cereals and grasses is similar.

1. The initiation of embryogenic callus of most cereals and grasses is dependent on the use of inflorescences containing caryopses at a specific stage after pollination as a source of explant tissue. Caryopses of rice 10–15 days after pollination are surface-sterilized then the embryos dissected out and placed with the embryo axis in contact with the nutrient medium in petri dishes (MS medium, 2–5 mg l^{-1} 2,4-D, 3% sucrose and 0.8% agar).

2. The petri dishes are sealed with Parafilm and maintained in the dark at 28°C when callus is initiated after 1–2 weeks. After 3–4

weeks, the callus appears to consist of two types easily distinguished by their color and appearance. The non-regenerating callus is translucent and uniform and shows a rapid growth, whereas the regenerant callus (Type I callus) is less homogenous and contains small densely cytoplasmic cells which will eventually give rise to embryos or shoots on the appropriate media.

3. The embryogenic callus is separated from the non-embryogenic callus and is subcultured every 2–4 weeks to ensure that the capacity for regeneration is maintained. However the proportion of non-embryogenic callus will increase despite separating the two types at each subculture.

4. The regenerating callus is transferred to half strength MS without growth regulators to stimulate the growth of the plantlets. When the shoots and roots are large enough the plantlets are transferred to soil.

7.3 Decline in totipotency in tissue cultures

Tissue cultures can be subcultured for many years without any visible change but examination of long term cultures shows the presence of increased levels of aneuploidy and polyploidy. Expression of somaclonal variation in regenerated plants is another indicator of internal change (Chapter 11). Cultures which are initiated and then maintained at high auxin levels in an undifferentiated state seem to gradually lose their capacity to re-differentiate with each successive subculture. This capacity can be measured by the number of embryos or shoot buds that the callus produces when placed onto a regeneration medium. This loss in totipotency may take place within ten subcultures after the original initiation of callus tissue. The reason for the decline in totipotency is thought to be due to a change in the cell population of the tissue culture. Although the loss of potentially meristematic cells can not be measured directly, the increase in cells showing aneuploidy and polyploidy with time is a good indication that deleterious changes are occurring in the cells. The fact that buds or embryos regenerated from this callus show much less obvious chromosomal variation suggests that the process of regeneration occurs from the more normal diploid cells. The proportion of potentially meristematic cells must decline in a subcultured callus and this is reflected in a reduced totipotency.

Although regenerating cultures seem to contain fewer chromosomal abnormalities, there is a concern that abnormalities do still accumulate. Thus the oil palm embryogenic cultures which provide a commercial source of tissue culture-derived plantlets, are only maintained for 10 subcultures.

During routine subculture of undifferentiated callus, there is always a tendency to select for pale-colored, vigorous tissue cultures, thus providing a selection pressure for cells adapted to the conditions of culture. By encouraging cell division this may accelerate the introduction of abnormalities into the culture. The various strategies for retaining totipotency all place a limit on the rate of cell division. These are that cultures are maintained at low auxin levels, under long subculture times and at reduced temperatures

The fact that there may be a selection pressure for cells that are better adapted to growth under *in vitro* conditions may explain why some tissue cultures, initiated from tissue culture-derived regenerated plants, show an increased regenerative capacity when used as explant tissue. For example in a recalcitrant commercial alfalfa variety, the number of regenerant buds on callus tissue derived from seedling explants was very poor. However callus derived from the regenerants showed a regeneration frequency seven times above that of the original callus. Seed progeny of the second generation of regenerants also showed the same enhanced regeneration capacity indicating that it was an inherited character. The presence of a higher rate of regeneration in the callus derived from the seed progeny suggested that the original meristematic cells that had given rise to the regenerant buds were altered in the tissue culture. The process by which regenerant individuals are used as a source of explants may be a strategy to enhance the regeneration frequency of normally recalcitrant genotypes (A. Safarnejad, personal communication).

7.4 Transfer of tissue culture-derived plants to soil

All tissue culture-derived plantlets have a poorly developed cuticle and often an inadequate root system. When transferred to soil there is rapid water loss and a risk that the plantlets will be lost or suffer severe setback. The method of overcoming this problem is to treat the plantlets as cuttings and transfer them to high-humidity conditions either in a room or in designated boxes for a period of 1 to 2 weeks before gradually hardening them (Chapter 13).

7.5 Markers for regenerating potential

Although now many species of crops can be re-differentiated from tissue culture, this has often only been successful with a restricted number of varieties for each crop. It is difficult to explain why there should be such marked differences in response to a tissue culture medium when the genotypes of the varieties are often so similar. The approach to the initiation of regeneration in tissue cultures of a new variety is to select a medium that has been successful previously with the same or related crop, then to modify it by screening for the optimum combination and concentration of growth regulators. In an effort to understand the process of re-differentiation, research has attempted to identify specific proteins and also mRNA sequences that are present only in potentially embryogenic cells or tissue. In addition the process of regeneration is being followed by identifying the sequence of changes in these characters during embryogenesis.

Genome expression is measured by the production of specific mRNA and protein from those genes associated with embryogenesis. An embryogenic gene from one of a group of genes (late embryogenesis abundant, *Lea*, genes) which are expressed in the later stages of zygotic embryogenesis was found to be expressed in somatic embryonic tissue and somatic embryos, but not in non-embryonic callus and young plantlets. Another feature of somatic embryogenesis is the wide array of secreted proteins that are released from tissue cultures and into the medium during the transition from non-embryogenesis to embryogenic cultures. These proteins such as glycoproteins, lipid transfer protein, acidic endochitinase, peroxidase and arabinogalactan proteins are closely involved in the developmental process. The immediate practical application of this knowledge is that these proteins which can be isolated from the medium surrounding the tissue may have the capacity to convert a nonembryogenic culture to an embryogenic state.

References

1. **Smith, S.M. and Street, H.E.** (1974) *Ann. Bot.* **38,** 223–241.
2. **Barwale, U.B., Kerns, H.R. and Widholm, J.M.** (1986) *Planta,* **167,** 473–481.
3. **Vasil, I.K. and Vasil, V.** (1990) In *Plant Tissue Culture Manual – Fundamentals and Applications* (K. Lindsey, ed). Kluwer Academic Publishers, Netherlands, pp. B1:1–161.

8 Haploid cultures

One of the aims of plant breeding is to improve crop species by the introduction of new variation. This is achieved by crossing between existing varieties of a crop, or between varieties of a crop and another species. Once the cross has been made and the F1 heterozygote produced, then it is necessary to eliminate the majority of the contribution from the paternal source except for the variation in a specific character(s). Elimination of unwanted genetic material is achieved by an extended period of further crossing and selection. The approach is dependent on whether the crop is an outbreeder or an inbreeder. In outbreeders, selected F2 (i.e. F1 × F1 in which segregation of the characters occurs) is back-crossed to the maternal parent, that is, the parent without the required variation, then the back-cross is selected for a combination of the required character and other desirable agricultural characteristics. In inbreeders, the F1 is allowed to self-pollinate to produce the F2 and the progeny selected for a combination of new variation and desirable characters. This cycle of back-crossing, or selfing, and selection will continue until the individuals of the cross are similar and are therefore considered to be homozygous. This process can take between 6–10 generations and as many years for annuals. With perennials the final selection occurs at the F1 or F2 stage. A program of backcrossing, or self pollination, and selection is out of the question for perennials because each generation may take 10–50 years before a selection can be made.

Homozygous individuals provide the source of commercial seeds since it can be guaranteed that all seeds within a variety are of a largely identical genotype and will produce similar plants. Homozygous varieties also provide the parents lines for the production of hybrid (F1) seed which are now used extensively in agriculture and horticulture. The seed from homozygous plants is therefore essential for plant breeding. The disadvantage of the traditional method of producing homozygous individuals is that the need for time, space, labour and ultimately cost is high.

The induction of dihaploids from haploid plants is an alternative route to the formation of homozygous lines. Haploids would normally be derived from the pollen or ovule of the original heterozygote (F1), and within one generation would produce homozygous progeny. In practise, in order to ensure that the characters are stable, the second generation progeny of the dihaploids are normally screened for the most desirable combinations. This still represents a considerable saving in time over traditional methods (*Figure 8.1*).

The source of dihaploid plants are the immature pollen or microspores which, at a specific stage of their development, are transferred aseptically to nutrient medium to initiate embryo and subsequently plant development. Ovules also provide haploid megaspores which can be used as a source of haploid tissue, and are treated in a similar way to the microspore. An alternative approach is to generate a hybridization between widely separated parents by pollination where it is known that the paternal chromosomes of one parent will abort in the first few divisions of the zygote. The embryo remains viable with the maternal chromosomes intact and is effectively haploid.

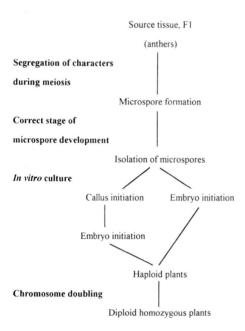

Figure 8.1. Sequence of stages in the production and regeneration of microspores.

8.1 Anther Culture

Although it is possible to obtain anther cultures from many species of crops and trees, the ability of any one species to produce microspores successfully is often restricted to one genotype or variety. The reason for this restricted response is unknown, and unfortunate, since the successful genotypes may not include the important commercial varieties. The choice of treatments to use with any new genotype or species can be established from the extensive background literature on anther culture in combination with the approach to regeneration described in Chapters 5–7.

8.1.1 Growth and development of the donor plant

For maximum embryo production from each anther, it is important to adopt the following:

1. The parent plants are watered and fertilized to maintain vegetative vigor. Flowers are selected at the correct developmental stage to provide microspores. *Figure 8.2* indicates the sequence of development and the most appropriate stages for anther culture. The critical stage varies from species to species; for instance in the Graminae and Brassicae it is the mononucleate stage, whereas in tobacco it is the stage prior to the first mitotic division.

2. The stages of development of the microspores in the unopened flower are first identified by cytological staining. Anthers are squashed and stained with acetocarmine solution (40% w/v carmine in 50% w/v glacial acetic acid, refluxed for 8 h, filtered and stored in the dark), then the stage of microspore development is correlated with the morphology of the flower or spike. The correct stage of microspore development is identified subsequently by using floral morphology as a guide.

8.1.2 Isolation of microspores

Once the stage of development of the microspore has been identified then its subsequent development *in vitro* can only occur if it is exposed to a particular pre-treatment.

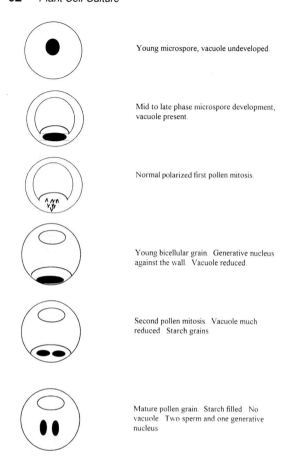

Young microspore, vacuole undeveloped.

Mid to late phase microspore development, vacuole present.

Normal polarized first pollen mitosis.

Young bicellular grain. Generative nucleus against the wall. Vacuole reduced.

Second pollen mitosis. Vacuole much reduced. Starch grains.

Mature pollen grain. Starch filled. No vacuole. Two sperm and one generative nucleus.

Figure 8.2. Stages in pollen development *in vivo* (Sunderland, 1974).

1. The pre-treatment varies from the use of short periods of high or low temperature, centrifugation, reduced atmospheric pressure, osmotic stress or N_2 starvation with no recognized or preferred pre-treatment. The Brassicae for example have a pre-treatment that involves excising the immature flowers then placing the cut stem in water and exposing the flower head to 1 day at 35°C, 1 day at 30°C then transfer to 25°C, whereas tobacco and cereals require a period of low temperature. In the Graminae a pre-treatment may not be essential. Excised flowers can in fact be stored for up to a week in the darkness at 4–7°C. For a new genotype or species, the choice is dependent on which pre-treatment was successful with other members of the family or near relations.

2. After the pre-treatment, the excised flowers are surface-sterilized before the flower is opened (Chapter 3). This part of the operation is performed in a sterile flow cabinet using a binocular microscope that has been surface-sterilized by wiping all surfaces carefully with 70% ethanol. Flowers are excised then surface-sterilized by immersion in 5% sodium hypochlorite (or commercial bleach) for 10 min. Once surface-sterilized, the floret is transferred to a sterile petri dish, then opened under the dissecting microscope, the anther is excised and large numbers transferred to nutrient medium in 9 cm petri dishes.

3. In cereals the unopened spike is surface-sterilized with 70% ethanol then the cut end is immersed in Nitsch and Nitsch nutrient medium in 100 ml conical flasks. The flasks are enclosed in sterile plastic bags then incubated at 26°C for 12 h in 1500 W m^{-2}. Single anthers are removed and stained in acetocarmine solution to establish the stage of development. When the pollen is at the first mitotic division, the flasks are transferred to a dark refrigerator at 6°C for 6–10 days. The anthers are then cultured on potato medium containing 0.25 mg l^{-1} 2,4-D.

4. Other conditions include the osmotic pressure of a medium which can be modified by the level of sucrose. Thus the Solanaceae require low osmotic pressures (2–4% sucrose) and the Graminae (9% mannose) and Cruciferae (8–12% sucrose) high osmotic pressures. Culture conditions are usually in the dark at 25–30°C. However once embryos are initiated, cultures are transferred to the light.

8.1.3 Initiation of embryos

The nutrient media used to initiate anther cultures are standard inorganic tissue culture media (Chapter 4, *Table 4.1*), or for the Graminae there are minor modifications of the basic MS medium (100 g l^{-1} KNO$_3$, 10 g l^{-1} (NH$_4$)$_2$SO$_4$, 20 g l^{-1} KH$_2$PO$_4$, 10 g l^{-1} CaNO$_3$•4H$_2$O, 12.5 g l^{-1} MgSO$_4$•7H$_2$O and 3.5 g l^{-1} KCl, 36.75 mg l^{-1} FeEDTA, 1 mg l^{-1} thiamine). The media may be liquid or solid, but it appears that there is less difficulty in regenerating plants from embryos on a solid medium. Some species show a requirement for growth substances, for example the Graminae, (0.5 mg l^{-1} kinetin and 0.5 mg l^{-1} 2,4-D) and Cruciferae (0.1 mg l^{-1} 2,4-D and NAA) show a requirement for growth regulators while those from the Solanacae appear to be independent. The concentration of growth regulators is critical as too high a concentration will cause callus formation. Examples of hormone combinations for the cereals are kinetin (0.5 mg l^{-1}) and 2,4-D (1.5 mg l^{-1}), or BAP (1.0 mg l^{-1} and IAA (1 mg l^{-1}). Although undefined media such as potato media have been used for cereals, it is better if cultures can be transferred to a defined media.

The pattern of regeneration varies with species and this will determine the nutrient media and culture conditions. Thus in the Solanaceae, the embryos grow directly out of the anthers and require transfer to a modified medium for further development into plants. In Brassicae the initial embryo development is followed by the formation of secondary embryos. These secondary embryos are then subcultured onto a hormone free medium for further development into plants. By contrast the cereals first give rise to callus, and the callus still attached or separate from the anthers is transferred to a medium with reduced sucrose and growth regulators (e.g. IAA 0.4 mg l^{-1} and BAP 0.4 mg l^{-1}) and the grasses (0.1 mg l^{-1} kinetin and IAA) to induce embryo formation. The cereals and grasses thus require a two stage process to initiate embryogenesis, and all species require a growth regulator free final stage to stimulate development of the embryos into plants.

8.2 Ovary and ovule culture

The culture of unpollinated ovaries or ovules has been attempted for many years but only recently has it been possible to obtain haploids of agriculturally important species via this route. Although viable haploid plants have been produced from sugar beet, wheat, tobacco, rice and maize, the overall yields of plants have been low. The technique is very labor intensive and is therefore of limited interest to the plant breeder. The only advantage that the technique might have is that for cereals the proportion of albino regenerants is significantly lower, and for sugar beet there is no alternative anther culture. The criteria for successful ovary culture in sugar beet, and other species, is dependent on similar pre-conditions as anther culture. These are that the growth and health of the donor plants is good, the correct developmental stage of the embryo sac is used, the flower buds and ovaries have the correct pre-treatment and the nutrient medium is optimal. As in anther culture there is a pronounced genotype affect. Studies on the sugar beet show that the system is less demanding than anther culture since the stage of development is less restricted and pre-culture treatments are not required.

8.3 Interspecific hybridization using excised embryos

An alternative approach to isolating and culturing haploids in order to obtain dihaploids is the 'bulbosum' technique [1]. Here there is a normal fertilization to produce a diploid zygote then one set of chromosomes is eliminated. The 'bulbosum' technique is used mainly in barley, and the paternal chromosomes are provided by *H. bulbosum*. Although early reports had indicated the cytological consequences of the cross, it was not until methods of embryo culture had been improved that the haploid embryo could be excised and grown into an intact plant. The barley, *H. vulgare* is fertilized by pollen of *H. bulbosum* causing a large proportion of zygotes to form. The chromosomes of *H. bulbosum* are rapidly eliminated from the cells of the developing embryo and loss of the endosperm also occurs. The embryo continues to grow although much more slowly than the normal diploid barley embryo. The embryo must then be dissected out of the fruit and transferred to nutrient medium if it is to survive.

8.3.1 Production of embryos

For optimum results the barley must be grown under the most favorable conditions and be disease free. *H. bulbosum* is a wild perennial which needs to be vernalized before it will flower. The procedure for making the cross is as follows:

1. Prior to hybridization, the barley florets are emasculated. Using fully mature flowers about 1 day before pollen release, the flower is cut across the top third avoiding damage to the stigma. This is done in the afternoon, and by the following morning the anthers will have begun to push through the cut opening and can be excised.

2. After emasculation, the spike is enclosed in a paper bag then by a cellophane bag which are then closed after pollination. The bags ensure a high level of humidity which is necessary for good fruit set.

3. Normally one *H. bulbosum* plant is required to pollinate 2–5 barley plants. Pollen is collected from freshly dehiscing spikes by tapping them gently. The pollen is collected on a camel hair brush and applied immediately to the stigma of the emasculated barley two days after emasculation of the barley anthers. Fruit set can

be improved by application of 75 mg l⁻¹ GA to the foliage 1–2 days after pollination.

8.3.2 *Culture of excised embryos*

Timing of the dissection of the embryo is dependent on differences in position of the fruit on the spike and genotype differences in fruit and embryo development.

1. The optimum stage for dissection of the embryo appears to be 13–15 days after pollination when grown at 23°C, and 18–21 days when grown at the lower temperature of 18°C.

2. At harvesting, the florets are removed then the outer layers, the palea and lemma, peeled off. The intact fruit are placed in a petri dish containing 5% sodium hypochlorite plus a drop of Tween-20 or -80 for 5 min, followed by five washes in distilled water.

3. The embryos are dissected from the fruit under magnification in a laminar flow cabinet working on the surface of a sterile petri dish. The fruit is slit open to release the liquid contents, then the embryo lifted out using a needle or fine tweezers, and it is placed on the surface of a sterile Millipore filter (GS 0.22 µm) on the surface of 8 ml nutrient medium (Gamborg B5) in a 5 cm petri dish and sealed with Parafilm.

4. The cultures are placed in the dark at 18°C for 1–2 weeks then moved to a 12 h light (1500 W m⁻²) regime and the temperature is raised to 20–22°C. When differentiated, the embryos are moved to a solid rooting medium (Chapter 5) after roots begin to develop in typically 1–2 weeks. Once plants are established they are transferred to higher light intensities and finally hardened off in soil under glasshouse conditions.

When plants have grown to the 3–4 tiller stage they are assessed for hybrids and barley monoploids. The hybrids are easily identified by their prostrate growth habit, slightly curled dark green leaves, and visible hair. There is also an upright growth habit, again with distinct hairs. Following this stage the chromosome number is doubled by colchicine treatment.

8.4 Doubling of chromosome numbers

The chromosome number of haploid tissue cultures or haploid plants must be doubled to provide dihaploid homozygous plants. Chromosome numbers of haploids will often double spontaneously during the tissue culture stage and, in the case of tobacco, if repeated cuttings are taken of the regenerated plants, spontaneous doubling of the dihaploid state will also occur. However, this method of chromosome doubling is unreliable, and to ensure that all the regenerants are doubled, the process is stimulated synthetically by exposure to colchicine. The colchicine is prepared fresh by adding 1 g colchicine (chloroform free) dissolved in 20 ml concentrated dimethylsulfoxide (DMSO) per liter of solution. Tween-20 or -80 is added as a wetting agent (0.2–0.5 ml l^{-1}).

Colchicine acts by inhibiting the spindle formation so that the chromosomes double up but do not separate as in normal mitosis. Once the colchicine is removed, the doubled-up chromosomes then undergo a normal mitosis with the doubled chromosomes appearing in all dividing cells. In the intact plant, colchicine treatment may lead to a hypersensitivity to excessive watering. Care must be taken so that colchicine treated plants are not waterlogged.

1. The doubling process is initiated in haploid tissue cultures by incubating the callus or embryos in the current liquid nutrient medium with colchicine added (e.g. 5×10^{-4} M) for 6–12 h. The callus is then transferred back to a regeneration medium.

2. Colchicine may also be used at the intact plant stage when it is injected into the axillary buds, or the shoot immersed in the colchicine solution (0.05%). In barley, for instance, at the 5 leaf stage, the complete shoot is immersed in this solution of colchicine in the light at 20–22°C for 5 h.

A comparison of dihaploids produced synthetically by colchicine treatment of tobacco tissue cultures derived from the anthers, and by spontaneous doubling by taking repeated cuttings *in vitro* from the haploid plants showed that there was no difference in the behavior of the haploids from each source.

Since colchicine is a toxic chemical, nitrous oxide (N_2O) has been tested as an alternative method of inducing chromosome doubling. Young wheat plants at the 3–4 tiller stage exposed to 6 atm of N_2O for

24–48 h showed a similar chromosome doubling response as immersion in 0.01% colchicine. Nitrous oxide can therefore provide a safer alternative to colchicine.

8.5 Comparison of methods of dihaploid production

The relative advantages of the three methods of dihaploid production can only be compared in the cereals since barley and wheat are the only crops on which the bulbosum technique is practised routinely. The two techniques that can be compared directly are the anther culture and 'bulbosum' techniques. Anther culture has the potential to produce large number of haploids with relative ease. The major disadvantage is the genotype dependency, and the high frequency of albinos. The bulbosum technique has by contrast already contributed to commercial varieties, but this also has genotype limitation and is restricted to barley and wheat. Since a plant breeder must choose varieties on agronomic grounds these limitations must be overcome.

The choice of which particular system of dihaploid production is used is determined by a number of factors. One of the most important is the frequency of dihaploids. In barley, where the comparison can be made, the 'bulbosum' technique was far superior. On average it produced 23.6 haploids per 100 florets pollinated, compared with 0.4 haploid plants per 100 anthers cultured. When a similar comparison was made with wheat on a per-spike basis, anther culture was slightly superior to the bulbosum technique in terms of embryo production. The other important criterion is the time required. In spring wheat, the time taken from sowing the F1 parent plant to harvesting doubled haploid grain on colchicine-treated haploids was almost 13 months for both haploid methods, and for winter genotypes it was 17 months. In comparison the conventional pedigree system would be at a similar stage by the F4 generation for spring genotypes and F3 for winter genotypes. This results in a time saving of 1–2 years for spring lines and 2–3 years for winter lines. The final criterion to be considered is cost. In view of the high capital cost of *in vitro* systems, only elite lines or those with a specific breeding objective would warrant the attention of a dihaploid system as against traditional methods.

8.6 Genetic variation in dihaploids

Normally the use of haploids would have a primary role in the production of homozygous lines. However, in cereals and grasses there is an intermediate callus phase before embryos are produced from anthers. This raises the possibility of a somaclonal type variation being introduced into microspore-derived plants (A. Zare, personal communication). This somaclonal, or gametoclonal, variation has been examined in barley by following the variation in dihaploid plants derived from homozygous lines compared to seed-derived progeny of the same lines. There was a greater variation in the anther-derived double haploids. Although the origin of the variation is not clear, it will contribute to the variation already generated in the microspores by the normal segregation and recombination between the parental genomes. In theory there is less chance of gametoclonal variation appearing in dihaploids from dicotyledenous species since there is no callus phase. However, field trials of dihaploid tobacco plants from homozygous lines showed variation in a range of characters which could only have been induced by mutation at an early stage in the tissue culture process. Dihaploid plants obtained via the 'bulbosum' technique show a degree of variation. It is thought that this variation might be caused by colchicine induced mutations.

The presence of gametoclonal variation in dihaploid plants provides a method of screening for useful variation *in vitro*. The approach would be to inoculate the anthers onto a medium which contain a specific compound that would select for resistant mutations. This additive could be salt, herbicide, heavy metals or phytotoxins. The embryos would therefore differentiate in the presence of the additives, and only those resistant cells would differentiate into embryos. Embryos produced by this *in vitro* screening could then undergo chromosome doubling and express any previously recessive resistant characters.

The anther culture technique has been applied successfully to a number of cereals and forage grass species (e.g. oats, rice, sugar cane, millet, barley, wheat and maize), oil seed rape and other Brassicae, (e.g. potato, tomato, tobacco) and other Solanaceae (vegetables, flowers, deciduous timber trees), with the overall list now extending to more than 200 species. Megaspore culture has been more limited but it has been successfully applied to sugar beet, wheat, tobacco, rice and maize. The 'bulbosum' technique, where the paternal chromosomes abort in the zygote, is of considerable importance but only to the cereals, barley and, to a lesser extent, wheat.

Reference

1. **Jensen, C.J.** (1977) In *Applied and Fundamental Aspects of Plant Cell, Tissue and Organ Culture* (eds J. Reinert and Y.P.S. Bajaj). Springer-Verlag, Heidelberg, pp. 299–340.

9 Protoplast cultures

The thick cellulose cell wall of the plant cell provides a protective and supportive function for the cell and ultimately for the plant. However the presence of the cell wall does mean that it is very difficult to transfer physically large molecules, vesicles or organelles into the cell. Such modification to the genome or cytoplasm requires that the wall is removed to give an isolated protoplast. The plasmalemma surrounding the protoplast can then be easily penetrated either by direct insertion of a syringe or by creating holes on its surface by electroporation. Under suitable conditions the protoplasts will regenerate a cell wall and divide repeatedly to form a callus. On an appropriate nutrient medium the callus will regenerate to give plants that contain modifications to the cytoplasm or genome (*Figure 9.1*).

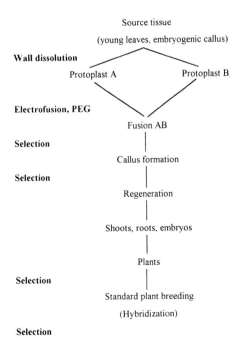

Figure 9.1. Stages in the isolation and fusion of plant protoplasts.

9.1 Protoplast isolation and culture

The technique was originally developed in dicotyledonous plants but has now been extended to many monocots [1]. The procedure can be divided into a number of stages. These are the selection of explants, pre-treatment of the explant, exposure to lytic enzyme solution and isolation and culture of the protoplasts. Each of these stages varies with the species and genotype but there are some general rules that can be applied.

9.1.1 Source of explant tissue

Cell walls of the explant tissue are removed by exposure to lytic enzymes then after treatment such as electroporation, or fusion, the protoplast must undergo cell division and regeneration. Consequently, explant tissue should be young and contain a high proportion of meristematic cells. This tissue consists of cells without excessive secondary thickening of the cell walls so that protoplasts can be released by only a short-term exposure to potentially damaging lytic enzymes. The high proportion of meristematic cells means that the tissue can respond quickly to a regeneration medium and is more likely to show cell division followed by embryo or shoot production. Source tissue is selected, maintained and treated as follows:

1. Source tissue in dicots is provided by young expanding leaves which must be surface-sterilized (Chapter 3) without damaging the tissue. An alternative is to surface-sterilize the seeds (Chapter 3) of the chosen species or variety, then grow the seedlings under *in vitro* conditions in a large container on MS inorganic medium plus 8% agar. When large enough the leaves and stem tissue are used.

2. A continuous supply of surface-sterile source tissue can be obtained by subculture of shoot tips of dicots. Shoot tips from surface-sterile seedlings are placed on a nutrient medium (0.1 MS medium and 1% sucrose). Once the shoot has expanded, the apical tip is removed to stimulate growth of the axillary buds. Terminal buds of the axillary shoots are excised and subcultured on the same medium so as to maintain a cycle of shoot tip excision and axillary bud growth. Leaf and stem tissue from the axillary shoots provides a supply of surface-sterile meristematic tissue.

3. It is difficult to regenerate plants from protoplasts isolated from leaf tissue of monocots. Immature embryos excised from fertilized

flowers produces embryogenic callus, which provides source tissue for protoplast isolation. The stage of development of the flowers is critical (e.g. 10–15 days after pollination). This tissue when cultured as a cell suspension retains the regenerative ability of the immature embryos and when isolated as protoplasts the cells are able to regenerate into plants.

Rice was the first cereal to be regenerated in this way and Orchard grass the first forage grass. Although a useful method for rice and subsequently maize, the other cereals showed a rapid loss of totipotency of embryogenic tissue cultures and require repeated initiation of tissue cultures from the source tissue.

9.1.2 Preplasmolysis and cell wall dissolution of explants

The media used for maintenance of tissue culture of the species or genotype being studied provides the basic nutrients for the culture of protoplasts. However, ammonium ions are often toxic to the protoplasts so this component could be reduced or omitted from the standard media. There may also be changes to the trace elements and organic components. The osmotic pressure of the incubating solution is particularly important. Since the protoplasts do not have a cell wall, the osmotic pressure of the external solution must be adjusted by the inclusion of non metabolizable sugars and sugar alcohols, such as mannitol or sorbitol. One of the most common of the isolation media for protoplasts is the CPW media (*Table 9.1*). The preplasmolysis and the lytic enzyme treatment of intact plants and tissue cultures is as follows.

1. Where the source tissue is leaves, stems or roots, it is cut into narrow strips and is maintained for one hour in the osmoticum solution which contains inorganic salt mixture such as CPW salts and 13% w/v mannitol as an osmoprotectant to prevent the cell from losing or gaining water after wall removal. If the source tissue is a tissue culture, the cells are broken up into smaller aggregates then maintained in the osmoticum at 25°C. This preliminary preplasmolysis incubation causes the cytoplasm to withdraw from the wall and makes wall removal less damaging.

2. The tissues are then exposed to the osmoticum, for example CPW plus 13% mannitol containing the lytic enzymes, either as a sequential or a mixed enzyme treatment. This incubation can be for a total of 2 h or overnight. The period of exposure depends on the source tissue and on the strength of the enzymes. For

Table 9.1. Composition of inorganic medium used for isolation of protoplasts

Constituents	Concentration in culture medium (mg l⁻¹)
KH_2PO_4	27.2
KNO_3	101.0
$CaCl_2\ 2H_2O$	1480.0
$MgSO_4\ 7H_2O$	246.0
KI	0.16
$CuSO_4\ 7H_2O$	0.025
pH	5.8

example, the sliced leaves of most species can be incubated in 1% cellulysin, 0.1% macerozyme R10 and CPW13M medium at pH 5.6 overnight in the dark. For cereals, where the leaves have been stripped of the epidermis, the mixture consists of 2% cellulysin, 0.2% mazerozyme R10, 0.5% hemicellulase, 1% potassium dextran sulphate, 11% mannitol at pH 5.8 using a Tris-malate buffer and incubated in the dark for 1–2 h and for most tissue cultures the incubation mixture consists of 2% rhozyme HP150, 2% meicelase, 0.03% macerozyme R10 and CPW13M medium at pH 5.8 and incubated in the dark overnight (see Appendix).

3. The cells are then freed of this mixture by filtering through a fine mesh filter (64 µm mesh size) and centrifuging at low speed. (100*g* for 10 min). The protoplasts are resuspended in 5 ml osmoticum without the lytic enzymes then the suspension is centrifuged as before. The pellet is separated from the debris by transferring by pipette to a dense solution (19–20% sucrose or 30% Percoll) then the centrifuging repeated which causes the debris to sink and the protoplasts to rise to the surface.

4. The protoplast band is then removed carefully by a pipette and transferred to a culture medium (MS medium, 2.0 mg l⁻¹ NAA, 0.5 mg l⁻¹, BAP 3% sucrose and 9% mannitol).

At this stage the protoplasts are at a suitable stage for fusion, or may be used for the uptake of large molecules such as in plant transformation (see Chapter 14).

9.1.3 Culture of isolated protoplasts

After isolation, the protoplasts are very fragile and still need an osmoprotectant until cell walls are formed. The media used for the maintenance of tissue culture of the species or genotype being studied usually provides the basic requirement but then the auxin and cytokinin levels may need adjustment. A typical medium would be MS medium 2.0 mg l⁻¹ NAA, 0.5 mg l⁻¹ BAP, 3% sucrose and 9% mannitol. The initial stage of wall formation and cell division to form a callus is the first barrier to be overcome.

1. The experimental approach to establish the correct auxin and cytokinin levels required to stimulate growth of the protoplasts is based on a Latin Square arrangement (Chapter 5). Where the protoplasts are restricted in numbers, an effective method is to use the hanging drop technique. Drops of media (100 μl) in which the cytokinin and auxin are varied are added to the under side of a plastic petri dish to give a Latin Square design. Known numbers of protoplasts (100) in an osmoticum are added as a small volume (100 μl) to the existing drops. This dilution of the auxin and cytokinin must be taken into account when preparing the initial concentrations. The petri dish lid is inverted over the base which contains sterile water to maintain humidity. The hanging drop technique can also be used with one concentration of medium to encourage cell division in a small number of protoplasts.

2. A method of stimulating cell division and regeneration that is technically easier is to transfer the protoplast directly to a liquid or semi-liquid medium in which the levels of auxin and cytokinin are varied. Because normal agar is toxic to protoplasts, it is replaced by low temperature gelling agaroses such as Sea Plaque, or Sigma types VII and IX. A suspension of protoplasts in the osmoticum is layered on the surface of the agar, or suspended in a small but equal volume volume of 1.2% agar at 40°C then the agar mixture (4 ml) transferred to 5 mm diameter petri dishes. Alternatively agar imbedded protoplasts can be incubated on a Millipore filter or nylon membrane over a nurse culture. The nurse culture is a rapidly dividing non-embryogenic culture from the same or a different species and also imbedded in agarose. The nurse callus is supposed to provide factors that stimulate cell wall growth and cell division in the protoplasts.

3. After two weeks the agarose block containing the colonies either on or imbedded in the agarose is transferred to fresh agarose medium in which the mannitol concentration is reduced to 6%. The colonies are transferred in this way to fresh medium every two weeks with a progressive 3% reduction in mannitol concentration.

The aim of all methods of protoplast culture is to protect the protoplasts from the loss of soluble cell components by restricting the volume of surrounding medium.

Once the callus is established and growing vigorously then it can be subcultured to a normal regeneration medium. If no established media is available for a new genotype then the procedures described in Chapter 6 must be adopted.

9.2 Methods of protoplast fusion

Isolated protoplasts from the same or different species can be fused under appropriate treatments, leading to a mixing of the nucleus and cytoplasmic material [2]. The heterokaryon is not necessarily stable and one consequence of the fusion is that there is sometimes a loss of one or more of one of the partners chromosomes and possibly even elimination of one set of chloroplasts. This process of chromosomal or organelle elimination can be controled by irradiation treatment of one protoplast partner which will destroy the chromosomes. Alternatively treatment of the cells of one partner with trichloracetic acid will destroy the organelle fraction but will leave the chromosomes intact. This means that it is possible to transfer the organelle fraction only into protoplasts lacking organelles so as to achieve a novel combination of cytoplasmic and nuclear components.

In order to encourage protoplasts to fuse it is essential that the plasmalemma membrane is destabilized then the protoplasts are brought into contact with one another. Protoplast fusion can be achieved either by chemical or by electrical methods.

9.2.1 Protoplast fusion using polyethylene glycol (PEG)

The simplest and least expensive method to achieve protoplast fusion is by chemical methods, of which the use of the polymer polyethyleneglycol (PEG) 6000 is the best known. The function of the PEG is to alter the membrane characteristics so that the protoplasts become sticky and if the protoplasts are allowed to come into contact they will adhere together and the contents will fuse.

1. The simplest combination to demonstrate protoplast fusion is to use a parent isolated from green leaves and one from colorless cell suspensions of the same (e.g. tobacco) or different species. In this way the parents and heterokaryon can be identified by their color.

2. Protoplasts isolated from each parent source to give a 4 ml suspension of protoplasts with a cell density of approximately 2 ×

10^5 in CPW plus 13% mannitol are mixed then centrifuged at 100g for 10 min so as to leave the protoplasts in approximately 0.5 ml of medium. This mixture need not be 1:1. For ease of identification of the heterokaryon between mesophyll and colorless protoplasts the ratio can be increased to 1:9 by altering the density of the suspensions.

3. The outer membranes are destabilized by adding 2 ml of PEG (30% w/v, PEG, mol.wt. 6000, 4% w/v sucrose, 0.01 M $CaCl_2 \bullet 6$ H_2O and autoclaved) and left for 10 min.

4. The protoplasts are fused by diluting the PEG every 5 min by adding a protoplast culture medium (MS medium, 2.0 mg l^{-1} NAA and 0.5 mg l^{-1} BAP, 3% sucrose and 9% mannitol) as increasing volumes (0.5, 1.0, 2.0, 3.0, 4.0 ml per tube). The protoplasts are resuspended after every dilution by gentle shaking.

5. The mixture is centrifuged at 100g for 10 min then the protoplasts are washed in the culture medium without PEG, centrifuged and resuspended in the same medium.

6. The protoplast suspension is then layered onto agarose medium or mixed with an equal volume of agarose medium, as described earlier, in order to stimulate wall formation and cell division.

The disadvantage of this method is that the PEG is often toxic to the plant cells and the percentage fusion can be as low as 1% which makes it difficult to retrieve the heterokaryons.

9.2.2 *Protoplast fusion using electrical fusion*

Protoplasts can now be fused on a large scale by electrical methods using chambers in which the cells are exposed to small electrical currents. Details of the procedure vary with the species and must be established from the literature, or empirically.

1. The suspension of protoplasts due for fusion is placed between two electrodes. A weak ac current (400 000 Hz, 1.5 V) is passed through the electrodes which causes the protoplasts to become positively charged on one side and negatively charged on the other. As a result of their charge, the protoplasts align themselves in groups along the lines of force, touching one another and forming pearl chains (*Figure 9.2*). The numbers in each chain can be altered by varying the protoplast density, the frequency of the ac field and the peak-to-peak voltage.

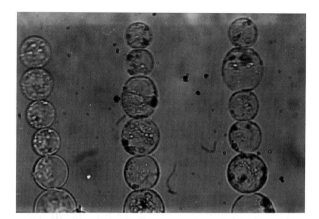

Figure 9.2. Pearl chains of celery protoplasts in resonse to a weak ac current.

2. After a period of about 90 s, the ac field is replaced by a single high voltage pulse discharge of for instance 1000 V cm^{-1}. Where the protoplasts are in contact, the plasma membranes break and fuse forming a continuous membrane around the pearl chains. This is then followed by cytoplasmic fusion (*Figure 9.3*). Multiple fusions will occur at high concentrations of protoplasts but this can be prevented by reducing the concentration of the protoplast suspension.

3. The protoplast suspension is then plated out as before.

In comparison, the electrical method of fusion is much more productive than the chemical method as up to 50% of the protoplasts may be as heterokaryons.

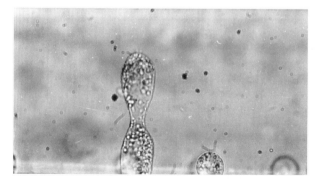

Figure 9.3. Cell fusion of celery protoplasts in response to a high voltage pulse.

9.3 Selection of heterokaryons

After both chemical and electrical fusion of the fusion mixture there will be single unfused parent protoplasts, fused parents of the same origin, multiple fusions as well as the desired heterokaryons. Isolation of the heterokaryons can be achieved by a number of techniques.

1. The simplest technique is to allow all of the protoplasts to regenerate and then to identify the heterokaryon at the seedling or mature plant stage by morphological differences. Where a large percentage of the protoplasts are heterokaryons as in the electrical fusion method, then simply regenerating all the protoplasts and identifying the heterokaryons at the intact plant stage is the most effective.

2. If only a small percentage of the protoplasts are heterokaryons then *in vitro* selection techniques are employed. A straightforward but laborious technique is where the heterokaryons are identified visually then removed manually using a microscope and micromanipulator. The suspension containing the heterokaryons is pipetted onto the surface of an agarose protoplast medium then the heterokaryons identified and transferred individually to fresh medium. This technique is suitable for the fusion of parent protoplasts with easily visual markers, such as in a fusion of green mesophyll and colorless tissue culture cells. Where the parents cannot be distinguished in this way, the original parent protoplasts are stained with single fluorescent stain such as fluorescein diacetate (FDA) which produces a green color or rhodamine isothiocyanate which produces a red color. The heterokaryon can be identified by its dual fluorescence in UV.

3. The other popular but more expensive method uses an automated flow cytometer. Here the dual labelled protoplasts are projected as a stream which is illuminated by UV. Those droplets containing the heterokaryon are identified and deflected into a separate tube. The procedure is fully automated and can process up to 2000 protoplasts an hour.

After regeneration of the isolated heterokaryon protoplasts, the heterokaryon status of the plants needs to be confirmed. This can be by morphological analysis since the plants should show a combination of parental characters, cytological analysis which will show whether there has been any chromosome loss during hybridization and biochemical analysis. The last would involve a comparison of the banding pattern of isoenzymes between the parents and hybrid plants, or genome analysis.

9.4 Contribution of protoplast fusion to crop breeding

Originally cell hybridization was seen as a means of bringing together two different genotypes that were unable to hybridize by normal sexual means. Very wide crosses were also thought possible by this means. However in the wide crosses there was often a loss of one set of chromosomes and also organelles. It was apparent that the partners had to be closely related for the heterokaryon to be stable. The best example of a successful protoplast fusion involving two related but sexually incompatible partners is that of the domestic potato, *Solanum tuberosum* with that of the wild potato, *S. brevidens*. The purpose of this cross was to introduce resistance to potato leaf roll virus and potato virus into the domestic potato from the wild relative. The chromosome number of the potato is 48 and that of *S. brevidens* is 24. In order to make the two partners more compatible, dihaploid plants were produced from *S. tuberosum* with a chromosome number of 24. Protoplasts from these plants were then fused with protoplasts from *S. brevidens*. The heterokaryons were regenerated and the plants were found to be tetraploids of 48 and hexaploids of 72. The latter were a result of three protoplasts fusing at the same time. When the hybrid plants were tested for resistance to both virus diseases, the hybrids showed the same levels of resistance as the *S. brevidens* and many were female fertile. This meant that they could be used as a source of compatible germplasm that could be crossed in a standard way with other potato varieties in order to introduce the virus resistance characters.

Another example of a successful cross which has potential for the production of hybrid seed is the induction of cytoplasmic male sterility. Where one partner in a cross is male sterile there will be no self-pollination in these flowers. In the past this possibility has been avoided by hand emasculation of the flowers which is an expensive and time consuming process. Any seeds obtained from crossing into a cytoplasmic male sterile female parent will be 100% hybrid seed. The expression of cytoplasmic sterility resides in the mitochondria and chloroplasts of the cytoplasm. Somatic hybridisation by protoplast fusion provides a mechanism for transferring the cytoplasm of a species with cytoplasmic male sterility to one that would benefit from this character. Rice is an example where this has been achieved.

Despite the early promises of the potential of protoplast fusion, it has not been possible to obtain stable heterokaryons of widely separated plant species. The other problem is that cell fusion is similar to normal plant hybridization so that there is a great deal of unwanted genetic material in the heterokaryon plant which has to be removed by the traditional methods of repeated backcross and selection. Of more interest to the plant breeder now is the use of specific gene transfer where only selected genetic material is incorporated into the crop species. Here plant protoplasts provide a method for the ready uptake of genomic material into the cell.

References

1. **Blackhall, N.W., Davey, M.R. and Power, J.B.** (1994a) In *Plant Cell Culture – A Practical Approach* (R.A. Dixon and R.A. Gonzales, eds). Oxford University Press, Oxford, pp. 27–39.
2. **Blackhall, N.W., Davey, M.R. and Power, J.B.** (1994b) In *Plant Cell Culture – A Practical Approach* (R.A. Dixon and R.A. Gonzales, eds). Oxford University Press, Oxford, pp. 41–48.

10 Preservation and cryopreservation of germplasm

Traditionally plant breeders have introduced specific characters into domestic cultivars by hybridizing with other current varieties, older varieties or wild species. There is an urgent need to provide long-term storage for this source of useful germplasm so that it is easily accessible for plant breeders. Samples of crop varieties and wild species are normally stored at low temperature as seeds, when they can be stored for a long period of time. In those instances where the plants are maintained as vegetatively propagated individuals, cuttings are taken at regular intervals and rooted so as to provide a continuing source of young plants. In plants that form storage organs, such as bulbs or tubers, clones are maintained by a continuous cycle of vegetative growth then bulb or tuber formation. This form of germplasm maintenance of vegetatively propagated species is expensive in labor and growing space and an alternative methods are always being sought.

There is the possibility that cell cultures may be a source of secondary products. This raises the question of how to maintain selected high-yielding cells without having to subculture them continuously. The standard method of maintaining *in vitro* grown material has been to subdivide the callus or cell suspension and to inoculate a small portion into fresh medium. When cultures are subcultured regularly over a long period of time, they accumulate somaclonal variation which may alter the pattern of secondary product synthesis. There is also the additional risk of introducing contaminants into the culture with a regular program of subculture. Reducing the frequency of the subculture, or stopping it altogether would reduce both the possibility of somaclonal variation and the risk of contamination.

There have been two approaches to the storage of vegetatively propagated germplasm. In one, the aim is to reduce the growth rate of the cells or plants and in the other, it is to stop the growth altogether. Both approaches are described in general terms here. For any one species and genotype the exact details of the technique will have to be established from the literature or empirically.

10.1 Reduction in growth rates of cells and tissues

Vegetatively propagated germplasms includes all those plants in which the germplasm is maintained by taking cuttings, or initiating tuber or bulb production or subculture of tissue cultures. The aim is to reduce the growth rate of the plants or cells and the frequency of establishing cuttings or subculture of tissue cultures.

10.1.1 Vegetatively propagated plants

In a crop such as the potato, breeding lines are maintained by a cycle of growth and tuber production. An alternative approach may be to isolate shoot tip cultures (Chapter 13), then maintain the cultures under a variety of restricted growth conditions. This type of approach could be taken with all of those species that are maintained by cuttings, or where elite individuals need to be preserved. In such *in vitro* methods, shoot tip cultures are placed in a reduced strength inorganic medium, such as 1/10 strength MS plus sucrose (1%), in order to limit growth. Alternatively growth is reduced by inclusion in the standard medium inhibitors such as 3–6% mannitol, or abscisic acid at 5–10 mg l^{-1}, or growth is inhibited by reducing the temperature of the tissue culture room to 5–10°C. The last is probably the least damaging, although care would need to be taken with tropical species which would have a higher minimum temperature.

When required, the cuttings are removed from the modified medium, transferred to a standard medium for 4 weeks, then rooted (Chapter 13), transferred to soil in the greenhouse and finally to the field.

10.1.2 Plant cell and organ cultures

For tissue cultures it is important to reduce the frequency of subculture of all types of tissue culture. In order to achieve this it is essential to extend the lag phase of growth of the culture and also to reduce the rate of growth within a subculture period.

1. At the beginning of a subculture, a reduction in inoculum size of cell suspension and callus culture will lead to an extension of the lag phase and will delay the onset of rapid cell growth. There is however a minimum inoculation density for cell suspensions below which the cells will not divide. These minimum inoculation densities will vary with the genotype and the medium and will have to be established empirically. There is not the same problem with callus subculture since callus cells are on solid medium and consequently suffer less from the effects of leaching. The smallest convenient inoculum for callus is 5–10 mg of tissue.

2. The rate of growth of tissue cultures can be reduced by lowering the temperature of the growth room to 10–15°C for temperate species and 15–20°C for tropical species.

3. The major problem with long term storage of all tissue cultures is the loss of water from both liquid and solid media. This can be reduced by ensuring that the seals of the containers are not excessively loose. If Parafilm is used as a seal then it may need to be replaced at intervals since it dries out and cracks.

Long term cultures may also show a higher rate of contamination. Care must be taken in checking for contamination when the cultures are due for subculture.

10.2 Inhibition of growth of cells and tissues by cryopreservation

The use of minimal media, low inoculum densities and reduced temperature will slow down growing material but the tissues still have to be subcultured regularly. The alternative to this strategy of reduced maintenance is to inhibit all growth by incubating the tissues at very low temperatures. At a temperature of −196°C (the temperature of liquid nitrogen), all plant growth stops. This is the basis of the cryopreservation method, which has been applied effectively to a large number of tissue types and species [1].

Cryopreservation was developed primarily for the preservation of cells and callus since these tissues were seen as a convenient way of storing all germplasm. Since then there has been an increased concern about the effects of somaclonal variation on the genetic composition of the preserved material, so that now there is more interest in the preservation of differentiated embryos and shoot tips.

Since cell suspensions are a potential source of secondary products, however, there is still an interest in cryopreservation of high yielding cell lines.

During cryopreservation, the cells or tissue are exposed to very low temperatures so that all activity is totally inhibited. Under these conditions the material can be maintained for a very long time then thawed and returned to normal conditions with all characters intact.

There are a number of very specific stages in the cryopreservation process. Because the treatment is so extreme, the tissue must first be conditioned before being exposed to the reduced temperature (pregrowth). The tissue is then exposed to −196°C (cryopreservation), then stored at this temperature (storage). When required it is thawed (thawing) and returned to normal conditions of culture (regrowth) (*Figure 10.1*).

10.2.1 Methods of cryopreservation

There are a number of methods that have been used on both cell suspensions and differentiated material. Except for the method used to cryopreserve cell suspensions, the methods for the differentiated material make use of relatively simple and inexpensive procedures. In

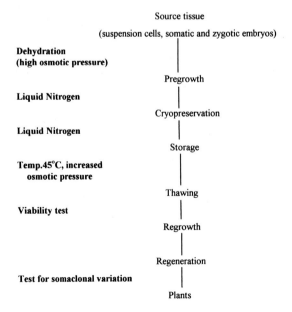

Figure 10.1. Sequence of stages in the cryopreservation and recovery of plant material.

order to perform a cryopreservation procedure, the following equipment and protective clothing are essential. These, are a source of liquid nitrogen, safety equipment (gloves, apron, face shield, pumps for dispensing liquid nitrogen from and to storage dewars), liquid nitrogen resistant dewars (these will provide vessels for transport of liquid nitrogen, for immersion of the material in liquid nitrogen during cryopreservation and for storage of the material in liquid nitrogen), and a programable freezer with dewar and pump. (This may not be required if differentiated tissues such as embryos or adventitious buds are used, which is increasingly the case now).

Cell suspensions. The most stress-resistant cell suspensions are those that form a friable fine suspension and consist of small densely cytoplasmic cells rather than large vacuolated cells. The suspension is more likely to consist of small cells if it is sampled during its growth stage.

1. Pretreatment. The cell suspension is inoculated at high density into standard medium containing 6% mannitol (w/v) and cultured under standard conditions. The suspension is harvested after 4 days when the cell will be dividing rapidly then concentrated up to 30% PCV. The suspension is chilled on ice.

2. Cryopreservation. A double strength cryoprotectant solution (1 M dimethylsulfoxide (DMSO) + 1 M glycerol + 2 M sucrose in culture medium) is chilled on ice. One volume of cryoprotectant solution is added to one volume of cell suspension and mixed with a magnetic stirrer in the cold room then the mixture is incubated on ice for 1 h. The mixture is dispensed as 1 ml aliquots into 2 ml capacity sterile ampoules. The ampoules are cooled at 1°C min^{-1} until they reach −35°C in the programable freezer, and are then maintained at this temperature for 30 min. Afterwards the ampoules are transferred to liquid nitrogen

3. Storage. The ampoules are stored in or over liquid nitrogen.

4. Thawing. The ampoules are dropped into sterile water at about 40°C with a ratio of 4 ampoules to 150 ml water. The water can be sterilized in a beaker with an aluminum foil cover which is left on while the ampoules are thawing. The liquid is agitated until the thawing is almost complete then the ampoules are transferred to a rack.

5. Regrowth. The cells in suspension are transferred to several layers of 5 cm filter paper on the surface of a 9 cm agar plate containing a growth medium and are incubated under standard

conditions for one day. The cells and upper layers of filter paper are then transferred daily to fresh medium, until after 5–6 days, the cells alone are transferred to agar medium.

Somatic embryos. Somatic embryos are an early stage in the differentiation process. Tissue in this form is genetically more stable than undifferentiated cell suspensions and after a period of cryopreservation will re-differentiate back to tissue cultures or differentiate directly into plants without going through a further period of tissue culture when somaclonal variation might occur. This may then be a more convenient way of storing tissue cultures

1. Pregrowth. Embryogenic tissue is cultured on standard agar medium containing 0.3 M sucrose for 2 months and is then transferred to standard agar medium containing 0.75 M sucrose for 7 days. The role of the sucrose is to progressively dehydrate the tissue so that it can survive the cryoprotectant treatment.

2. Cryopreservation and storage. The embryogenic tissue is located more specifically (often as localized green areas) then dissected out under a binocular microscope and transferred to plastic ampoules and immersed directly in liquid nitrogen and stored under liquid nitrogen.

3. Thawing. The ampoules are thawed at 45°C then using the filter paper technique are transferred to standard agar medium containing 0.3 M sucrose.

4. Regrowth. After 2 weeks the embryos are transferred to standard agar medium containing 0.1 M sucrose. After a further 2 weeks the embryos are transferred to an agar regeneration medium to stimulate plant formation.

Zygotic embryos. A number of seeds of tropical species are difficult to store because of their high water content when mature. An alternative is to remove the immature or mature embryo and cryopreserve this tissue. The details of the drying procedure would have to be worked out for each species.

1. Pregrowth. The embryos are excised aseptically from the seeds, then in an open petri dish, are exposed to a sterile airflow in a laminar flow cabinet for 3 h.

2. Cryopreservation. The embryos contained in plastic ampoules are immersed in liquid nitrogen.

3. Thawing. The plastic ampoules are transferred to a water bath at 37–38°C then placed on moist sterile filter paper in petri dishes for 10 days. No medium is added at this stage.

4. Regrowth. After 10 days in the absence of medium the embryos are added to a standard growth medium to stimulate growth into plants.

Adventitious buds. Stems contain adventitious buds in the axils of each leaf. These present a valuable source of potential shoots. The advantage is that the buds are small, protected and because they are dormant may already show the low hydration that would be required for a successful cryopreservation. They do not require a sophisticated pre-treatment which makes the cryopreservation simple and inexpensive.

1. Pregrowth. Uniform nodes (5 mm in length) containing an adventitious bud are removed from stem sections of plants. Nodes are placed on agar medium containing 0.7 M sucrose and incubated under standard conditions for 2 days then transferred to a nylon membrane contained in a dish with 15 g dry silica, sealed with Parafilm and desiccated for 4–8 h.

2. Cryopreservation. Nodes are transferred to plastic ampoules then immersed in liquid nitrogen and stored under liquid nitrogen.

3. Thawing. Nodes are thawed in a water bath at 25°C.

4. Regrowth. Because of their size, nodes can be transferred individually to a standard growth medium and the adventitious bud is stimulated to develop as a shoot. When large enough the shoot is rooted.

10.3 Confirmation of viability and stability of cryopreserved tissues

It is important to confirm that the cells or differentiated tissues have survived the cryopreservation procedure. Absence of growth in the regrowth stage will indicate that the conditions have been too harsh and the plants have not survived. However, a more rapid method of indicating the viability of the tissue is to use a viability stain. The most commonly used is fluorescein diacetate (0.1% FDA in acetone). A few drops of the stock solution of FDA is added to the cells or thin section of tissue and examined under a UV fluorescence microscope using a blue/violet filter. The presence and number of fluorescing cells in dark field is counted and compared with the total number of cells in bright field. This will indicate the proportion of live cells and hence the efficiency of the cryopreservation procedure. Another viable stain is Evans Blue (0.05 g Evans Blue/100 ml water or medium). The cells or tissue sections are incubated in the solution then examined for the entry of stain into the cell. The absence of blue color in the cells indicates viability.

The stability of the cells and regenerated tissue is established by genomic analysis using molecular biology techniques such as randomly amplified polymorphic DNA (RAPDs) to indicate whether there has been any genetic change during the storage period. This will apply particularly to undifferentiated callus and cell suspension and less so to regenerated tissue. For a further assessment of possible variation induced in regenerated tissue, it is important to allow the cryopreserved shoot tips or embryos to grow into plants then to assess the level of variation by morphological analysis. This morphological variation may be nongenetic in which case it can be screened out by a cycle of vegetative propagation or seed formation (Chapter 7).

Reference

1.   **Benson, E.E.** (1994) In *Plant Cell Culture – A Practical Approach* (R.A. Dixon and R.A. Gonzales, eds). Oxford University Press, Oxford, pp. 148–167.

11 Selection for somaclonal variants

Plant tissue cultures isolated from a single explant, or even from a single cell after repeated subculture will show variation in morphological characters such as growth rate, color, friability and regenerative ability and physiological characters. It is possible to select distinct lines from this one source with their own particular morphology and physiology. This suggests that the tissue cultures are not a mass of identical cells but a population of different genotypes whose proportion can be altered by imposing an appropriate selection pressure. The presence of such variation in cell type in callus or suspension culture forms the basis of *in vitro* selection procedures where cultures can be screened and cells with a particular characteristic isolated. In this way cell lines can be obtained which show increased herbicide resistance, or enhanced capacity to synthesize secondary products (*Figure 11.1*).

This variation in cell type can be transmitted to plants regenerated from the culture. The regenerants will often show considerable variation in morphology such as height, leaf number, leaf size, color and also in such physiological characters as resistance to heavy metals and disease. If the plants are then allowed to set seed and the first and subsequent generations of progeny are examined, some of this variation is retained. This variation in tissue culture and in regenerants and their progeny is described as somaclonal variation. It is regarded as a novel source of variation for the plant breeder, and for the industrial biochemist attempting to enhance the yield of specific plant secondary products in large-scale plant cell culture. There has been much effort devoted to identifying the source of somaclonal variation, the cytological and molecular changes, the methods used for the selection of specific characters and the stability of these characters in long-term cultures and in regenerated plants and their progeny.

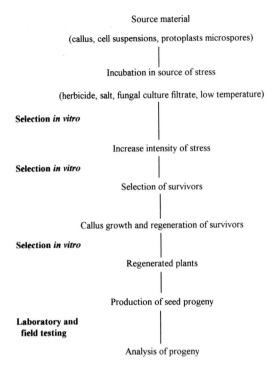

Source material

(callus, cell suspensions, protoplasts microspores)

|

Incubation in source of stress

(herbicide, salt, fungal culture filtrate, low temperature)

Selection *in vitro*

|

Increase intensity of stress

Selection *in vitro*

|

Selection of survivors

|

Callus growth and regeneration of survivors

Selection *in vitro*

|

Regenerated plants

|

Production of seed progeny

**Laboratory and
field testing**

|

Analysis of progeny

Figure 11.1. Sequence of stages in cell selection *in vitro* for somaclones resistant to biotic and abiotic stress.

11.1 Source of somaclonal variation

Somaclonal variation may be derived from variation present in the original parent explants, as well as variation produced during the initiation and maintenance of the tissue cultures.

11.1.1 Parent plant

It has been proposed that plants, especially perennials, are genetic mosaics that result from the production of somatic mutations in the meristem and cambium throughout the life of the plant. Where these mutations are non-lethal and continue to be expressed by dividing cells in the meristem and cambium, they can result in a permanent modification to sites of cell division and cell differentiation and ultimately to the pattern of organ formation. Thus, explants taken from the same plant will give rise to differences in the tissue cultures and these differences will be passed to the regenerant plants. The contribution of variation between and within plants to the incidence of variation in tissue cultures is shown by two examples of somaclonal variation in the regenerants of four varieties of *Medicago sativa*, there was a variation of 10–64% aneuploidy depending on the varietal source. In the potato, tuber explants produced regenerants where over 50% of the plants were aneuploid, whereas regenerants from leaf, stem and root were unaffected. In pea, callus cell derived from leaf explants showed 90% diploidy whereas in stem only 70% were diploid and root callus 50% were diploid [1].

11.1.2 Tissue culture initiation and subculture

In tissue cultures, new variation appears to be introduced by the initiation and maintenance of the tissue culture itself. An important influence is the presence of growth regulators such as 2,4-D. In a comparison of the effects of growth regulators in the medium on tissue cultures of *Kalanchoe blossfeldianae,* only 2,4-D caused high levels of morphological abnormalities in the regenerants. Here the effect of growth regulators may be a direct one on the incidence of somatic mutations, or a stimulus to the proliferation of specific cell types. As might be expected the longer the exposure of the callus phase to such influences the greater chance there is of increased variability. Examination of long term cultures shows a very high incidence of chromosomal abnormalities. The suggestion is that such changes occur during the first few cell divisions in the explant. In maize, for example, identical abnormalities were present in plants regenerated from callus at various stages of subculture. The later period of cell division during callus growth and subculture merely altered the proportion of the mutations in the culture.

Although the callus phase is very critical for transmitting variation from the explant, and also initiating it in the culture, the regeneration stage can act as a very effective screen. Thus in the Graminae, somaclonal variation is predominantly associated with regeneration from unorganized cell cultures, whereas those regenerants from embryogenic cell cultures were largely variation free. In addition, in tetraploid alfalfa, the embryogenic process also appears to act as a screen since globular embryos showed little chromosome aberration despite a high proportion of abnormal cells in the callus. However, the selective role of embryogenesis does not apply to all species. Somaclonal variation occurred in regenerants from somatic embryos of celery, and in soybean the frequency of genetic and cytogenetic aberrations was actually higher in embryogenic than organogenic cultures. Some species appear able to tolerate a high level of variation, whereas a species such as barley does not. Non-regenerative callus of kales contained high frequencies of aneuploidy, haploidy and tetraploidy whilst morphogenetic callus was predominantly diploid, whereas by contrast immature embryos of rye were found to contain aneuploids and tetraploids [1].

11.1.3 Chromosomal and molecular basis of variation

The cytological changes that can occur during tissue culture infection have already been mentioned. These changes are thought to arise from nuclear fragmentation or amitosis during the initial cell division. Mitosis then transmits these changes throughout the culture. Tissue culture initiation also leads to chromosome breakage of late replicating DNA (blocks of condensed heterochromatin that are replicated very late in the cell division cycle). This leads to the formation of heterochromatin bridges which can break and refuse thus causing rearrangement of the chromosomes and the induction of variation.

The molecular basis of somaclonal variation is based on the use of restriction fragment length polymorphisms (RFLPs) which established that alterations were due to changes in the DNA sequence and in methylation. One factor that can cause changes in the DNA sequence is the activation of the transposable element. Transposable element activity has been detected in the progeny of maize plants, whereas the activity was not present before culture. The other mechanism for heritable change is by gene inactivation by DNA methylation at position 5 of the cytosine nucleotide. These changes in methylation occur early on within the first few divisions of the initiation of leaf callus in Petunia hybrids and are maintained subsequently during future generations. The role of DNA methylation in somaclonal variation has yet to be clarified.

11.1.4 Control of somaclonal variation

It is important to control somaclonal variation so as to be able to either repress or enhance it depending on the purpose of the tissue culture. The determining factors appear to be the degree of somatic variation in the original explant and the level of organization of the tissue culture. The occurrence of somaclonal variation can be minimized by isolating explants from non-polysomic species (i.e. cells that are primarily diploid), by isolating explants from as close to the meristem as possible, use of younger tissue and essentially undifferentiated structures such as leaves. In this way the contribution of somatic variation from the parent tissue is kept as low as possible. At the tissue culture level, it has been found that the higher the level of organization in the tissue culture, the lower the level of variation. Thus maximum variation is detected in protoplast-derived cultures, and least in meristem or shoot tip cultures. A minimal period of culture will also reduce the proportion of aberrant cells, since tissue culture initiation appears to disrupt the cell cycle, possibly because the cells are dividing too fast and cause chromosome fragmentation. A reduction in the rate of cell division such as by a lowering of temperature, or an alteration of the growth regulator supply, may reduce these tissue culture-induced changes. Where it is required to increase the proportion of somaclones these limitations can be removed.

11.2 Selection for somaclones

The approach to screening for somaclones is dependent on whether the process is undertaken *in vivo* or *in vitro*. The advantage of producing regenerated somaclones first, then carrying out an *in vivo* screening procedure, is that the selection is imposed directly on the plants and therefore there is less likelihood of obtaining inappropriate mutations. The disadvantage is that an *in vivo* screen involves very large numbers of plants, which may be difficult to accommodate.

The alternative is to screen *in vitro* when very large numbers of individuals can be screened very conveniently and only small numbers of regenerated plants are produced as survivors. The selection methods used are based on selection for resistance, visual selection and chemical selection. In resistance selection, the aim is to increase resistance to disease, heavy metals, salt, herbicides or chilling in a specific genotype. In most examples the purpose of the selection program is to improve crop resistance. However, resistance selection may also be used to identify protoplast-derived hybrids and transformed cells which carry the resistant trait. Where the selected genotype is colored then a visual selection can be practised. In this way single cells with enhanced anthocyanin can be identified and isolated to establish high anthocyanin-producing cell lines. Some of the manual methods for isolating cells with high levels of alkaloids are based on the presence of UV fluorescence and should be described as a form of visual selection.

In most examples of selection not based on resistance, the preferred genotype is not usually colored so other methods of identification must be used. Under this category are populations screened for auxotrophs. Examples of such auxotrophs are amino acid and nitrate reductase deficient auxotrophs. Alternatively individual cells or cell aggregates producing enhanced levels of secondary products are isolated from the rest of the culture by chemical or immunological tests and multiplied to produce high yielding cell lines (see Chapter 12).

11.2.1 In vivo *screening*

The most noticeable feature of plants that have been regenerated directly from plant tissue cultures without any attempt at *in vitro* screening is the wide range of morphological variation (*Figure 11.2 a*). This includes intensity of chlorophyll, leaf shape and size, height, branching pattern and flower number, shape, and color. There are also likely to be physiological changes such as disease resistance and herbicide resistance. The fact that the changes are maintained to the stage of flowering in annuals is an indication of the permanency of the expression. The source of these wide changes is uncertain. Most are non-genomic but the changes are expressed throughout the life of the annual plant and for many years in the perennial. It is therefore of great interest to those attempting to manipulate the expression of variation in both herbaceous and woody perennials.

Figure 11.2a. Leaf shape of carrot plants regenerated directly from callus cultures.

Figure 11.2b. Abnormal shoot tips of Brussels sprout plant derived directly from callus.

To test the permanent nature of the changes in regenerants of perennial plants it is important to take cuttings from the selected regenerant individuals. This will provide a large amount of uniform material with which to use in a large scale replicated test. It is unadvisable to carry out the screen directly on the regenerants in view of their extreme variability. Vegetative propagation of the selected regenerant will also indicate whether the character was stable during vegetative propagation. In perennials some reversal to the parent morphology of the regenerants can occur over a number of years. Thus, the gross changes to plant and fruit shape that have been found in some of the tissue culture derived oil palm trees have shown a reversal as the plants age.

Although a wide range of variability appears in the regenerant generation, this variation largely disappears in the progeny of the first seed generation. Production of the seed seems to act like a screen on the variation of the seed generation. Only genomic changes will be transmitted to the pollen and ovule and in those individuals where large genomic and chromosomal changes have occurred pollen and ovule production will not take place. Seed formation acts like a screen so that only the small and least damaging mutations are transmitted and appear in the progeny. These seeds provide a source of uniform material with which to produce plants to test for the expression of stable changes in morphology and physiology under well-replicated conditions.

11.2.2 **In vitro** *screening*

The major application of *in vitro* screening is in the selection of resistant mutants which will provide the variation required for a crop improvement program [2]. The assumption is made in the selection at the cell level that the resistance to a particular set of cultural or environmental conditions is expressed in the same way in the cells as in the intact plant. This assumption may not be true for all resistant mutants. Drought tolerance may have as much to do with the length of the roots of the intact plant as metabolic resistance to drought. Cell selection cannot identify the ability to produce longer roots. This limitation of *in vitro* selection must be borne in mind. The smallest, and least organized units of selection are protoplasts, then in order of increasing complexity there are cell suspensions, callus and differentiated tissue. The strategy for producing resistant mutants is essentially the same for all types of tissue culture.

1. Initially the culture is placed on or in a normal medium which contains the selective agent (salt, herbicide, heavy metal or phytotoxin) at a broad range of concentrations. The phytotoxin is isolated from axenic liquid cultures of the pathogenic fungi or bacteria then either purified or included in the medium as a crude concentrated filtrate. The initial test is designed to establish the concentration which kills over 90% of the culture (D_{90}).

2. Having identified the D_{90} concentration, further batches of tissue culture are then exposed to this concentration in the nutrient medium.

3. Surviving cultures are identified then multiplied on a normal medium and finally regenerated.

4. Alternatively, the culture is exposed to a stepwise increase in the concentration of the additive and the surviving cultures transferred to the next concentration until a ceiling concentration is reached. The surviving culture is bulked and regenerated as before. In order to prevent the culture becoming diluted with each step, the period in each subculture is progressively extended so as to ensure the final number of cells in the stationary phase is always similar. The inoculum size is therefore constant.

There are small variations in the overall strategy depending on the nature of the tissue culture system. Details of these variations are as follows:

Protoplasts. The regeneration stage is a problem with any tissue source. So as to ensure that it is less of a problem with protoplasts, isolation of protoplasts is always made from intact plant tissue rather than existing tissue cultures. The main advantage of the protoplast system is that they are single cells so that there is no danger of chimeras being formed. It also means that it is realistic to expose the protoplast to a mutagen since each cell will be exposed uniformly. An example is provided by mesophyll protoplasts of *Nicotiana plumbaginifolia* which were exposed to the mutagen N-ethyl-N-nitrosourea and subsequently, callus colonies derived from the protoplasts were plated on to a medium containing triazine herbicide. Selection was based on the presence of chlorophyll in the resistant colonies.

Cell suspension cultures. Ideally a cell suspension should consist of a mass of single cells and small cell aggregates. The stepwise approach to selection is more widely used with suspension cultures. For example, cell cultures of *Nicotiana tabaccum* were subcultured in a stepwise procedure from 10 g l^{-1} to 35 g l^{-1} (599 mM NaCl) with the cells maintained for 50 generations in each medium since this allowed the cells to be bulked for the next inoculum. This approach has been used to produce glyphosate tolerance (25 mM) in *Daucus carota* and cadmium tolerance (5000 mM) in *Lycopersicon esculentum*. The most serious disadvantage of cell suspension cultures is the difficulty in regenerating the culture after such a prolonged period as a cell suspension. In the absence of regeneration the stability of the cell line is normally confirmed by maintaining the cells on a additive-free medium for at least 30 subcultures, then retesting. Thus glyphosate resistance in *Daucus carota* was still present after 120 subcultures [2].

An alternative method using cell suspension culture is to spread the cell suspension culture on the surface of an agar medium containing the specific additive. This allowed for the selection of zinc resistant cells in *Haplopappus gracilus*. A refinement to improve contact between the cells and the additive is to suspend the cells in liquid medium containing the additive for 24 h then to plate the suspension on agar medium which also contains the additive. Thus, cells of *N. tobaccum* were exposed to 1 mM glyphosate in such a two-stage process. The advantage of this approach is that by plating out the suspension, the results are much more visual. In addition the period of time as a cell suspension is much briefer and the chance of regenerating the selected cells therefore much improved.

Callus cultures. Callus cultures have been used routinely in selection for resistance to additives in the medium. The risk that not all the cells in the callus may be uniformly exposed to the additive is reduced by using very small pieces of callus. Cells resistant to the additive in the medium then grow out as a small mass on the side of an otherwise dead callus piece. The new growth can be easily identified and subcultured. The risk of producing chimeras is reduced by exposing a large number of very small pieces of callus to the selection such as 50 × 15 mg pieces of callus in a 9 cm diameter petri dish. In further screens for salt tolerance small callus pieces of 20–150 mg can be used. Selection involves either a single stage or a stepwise approach. In the single stage it has been possible to select for resistance to salt, amino acid analogues, herbicides and toxic or heavy metals. The stepwise approach has allowed the selection of, for instance, salt tolerance in *Coleus blumeii*, from 30 mM NaCl to 150 mM with a stepwise increase of 15 mM every 3 weeks, and *Beta vulgaris* in a similar method of increase showed tolerance to 210 mM NaCl in the medium. Providing the cultures are recently initiated it is possible to regenerate selected cultures and test for resistance in the progeny.

Using these *in vitro* selection methods, salt tolerance was detected in the progeny of plants regenerated from the selection program of *Coleus blumeii* and flax.

Differentiated cultures. The purpose of many *in vitro* screening programs is to produce plants that can be used as a source of variation in a breeding program. A constant problem in any *in vitro* selection is that the potential for differentiation may be lost in the period of tissue culture before and during screening. This problem can be avoided by carrying out the screening on differentiating material such as embryogenic or organogenic cultures. Initiation of embryos or shoots is thought to occur from single cells so that although the selection system involves multicellular structures such as embryos, their origin is from a single cell. The limitation of this approach is that not all cultures can produce large numbers of embryos or plantlets. However in those examples where differentiation is prolific, such as embryo production in celery or carrot or organogenesis in tobacco, such differentiating cultures can be used to screen for resistant mutants. Large numbers of celery embryos have been screened for herbicide resistance in culture and this resistance was expressed in mature regenerated plants.

11.3 Progeny testing of somaclones

In vitro selection will result in a cell line with specific resistant characteristics, or the ability to produce high yields of secondary products. Where the selection has been based on long-term tissue culture it may not be possible to regenerate plants. Stability of the selected character is shown by maintaining the selected cell line in the absence of the selection pressure, that is, in the absence of, for example, salt or herbicide for a large number of subcultures. The cell line is then retested for resistance. Experience has shown that cultures selected for high yields of secondary products are not always stable and the average yield of the culture may decline in the absence of the selection pressure.

Where a regenerant plant is required as part of a crop improvement program then the regenerant can be tested to see whether the character, although selected at the cell level, is expressed in the intact plant. It may not be very satisfactory to test the regenerant particularly if numbers are limited. The regenerant plant can then simply act as a source for seed production. The seed is then used in a progeny test for the selected character. Progeny testing of the first and subsequent generations of progeny is essential to establish the heritability and stability of the selected character. Only when the character can be stably transmitted to the progeny can the somaclonal variant be included in a plant breeding program.

11.4 Application of somaclonal variation to crop improvement

The early investigations suggested that somaclonal variation could make a significant contribution to crop breeding. The regenerant generation always shows a wide range of variation in morphology but much of this is lost in the first seed progeny. Although the variation does affect all characters, and not always those that are agriculturally useful, by selection it has been possible to produce useful lines from this source. For example, selection has produced a commercial line producing a high solid tomato, increased resistance to the herbicide chlorosulfuran in field grown maize, 10–100 fold increase in tolerance to the imidazilinone herbicides in maize and resistance to *Helminthosporium sativum in* wheat and barley and salt tolerance in flax. There has also been improved freezing tolerance, grain quality and protein content in wheat, and heavier seeds with higher protein content in rice. Although there are a number of examples of successful application of somaclonal variation to crop breeding, significant advances are still being made in the major crops by conventional techniques. The increased application of somaclonal techniques to breeding might be through new methods of crop improvement [3].

References

1. **Karp, A. and Bright, S.W.J.** (1985) *Oxford Series Plant Mol. Cell Biol.* **2,** 199–234.
2. **Collin, H.A. and Dix, P.** (1990) In *Plant Cell Line Selection* (ed. P. Dix). Springer Verlag, Heidelberg, pp. 3–18.
3. **Larkin, P.J. and Banks, P.M.** (1985) In *Current Issues in Plant Molecular and Cellular Biology* (eds M. Terzi, R. Cella and A. Falavigna). Kluwer Academic Publishers, Netherlands, pp. 225–234.

12 Secondary product synthesis by plant tissue cultures

The term secondary products covers a wide range of complex organic compounds that are synthesized in plants. Secondary compounds, which include as their major groups, the alkaloids, phenolics, flavonoids, steroids and terpinoids are often characteristic of a plant family. Thus celery (*Apium graveolens*) a member of the Umbelliferae contains significant amounts of terpinoids (e.g. pinene, limonene and caryophyllene) and polyketides, and *Atropa belladonna* of the Solanaceae contains the tropane alkaloids, hyoscine and hyoscyamine. Differences in structure of the secondary products and their sites of location in the plant make it difficult to identify a common function for all secondary compounds. There is evidence for them having a very important role as constitutive compounds in the plant's defense against insect pests and predators, particularly in mature tissue which accumulates these compounds. Secondary compounds can also be synthesized rapidly in response to microbial infection and insect attack, and the same compounds appear to be synthesized in response to a range of abiotic stresses such as cold, heat, osmotic extremes and physical damage.

Synthesis of secondary products appears to be stimulated if fixed carbon is not fully utilized by the primary metabolic activities of cell growth and differentiation. Carbon not used to synthesize cell walls and protein is converted into secondary compounds and stored in the vacuole or cytoplasmic vesicles. When more rapid growth is resumed then the secondary products are degraded and the stored carbon released. The activities of secondary and primary metabolism are therefore closely related and exist in a dynamic equilibrium linked by key enzymes. Phenyl ammonia lyase, for example, provides the link between intermediary metabolism and phenyl propanoid pathways. These key enzymes are very important as their activity may be a major rate-limiting factor in secondary product accumulation.

Secondary compounds have provided an array of complex structures to identify and describe but, because of the complexity of the biosynthetic pathways, knowledge of their synthesis has lagged behind. Current interest in manipulating the levels of these compounds in plants and tissue cultures has stimulated an increased interest in their synthesis.

Secondary products occupy an important place in the food, cosmetic and pharmaceutical industries. The taste of vegetables and fruit is a consequence of different flavor components which may be many, or limited to a few compounds. These same compounds are often added as an extract to many convenience foods. Food flavor is also enhanced by adding a very wide variety of herbs and spices, all of which contain specific secondary products. The flavor sources in tea, coffee and cocoa, are numerous and vary with plant source and method of preparation, whereas the stimulating qualities are related to the presence of the alkaloids, theobromine, theophylline and caffeine. Flavor in alcoholic drinks may be as specific additives, such as in hops, or the compounds present in the original plant source. These latter flavors and colorings are very numerous and can vary considerably depending on the source tissue, climate and preparation. Plants also contain a variety of essential oils which form the basis of the perfume and cosmetics industry and provide additives for domestic and industrial detergents. The most valuable plant secondary products are those used by the pharmaceutical industry for the treatment of cancer (vinblastine, vincristine and taxol), as anticholergens (atropine) and as heart stimulants (digoxin). Even now, 25% of all pharmaceutical products are plant-based. Plant secondary products make a contribution to the well being of peoples lives and as such provide a considerable revenue for the industries concerned.

Many of the plants which provide the herbs, spices, beverages, perfumes, essential oils and pharmaceuticals are grown in areas of the world where production may be limited by climatic conditions, or the unsophisticated nature of the agriculture. The possibility of an uncertain and variable supply of the products has prompted an examination of alternative and possibly more stable sources. One possibility was to treat plant cell suspensions in the same way as microbial-based fermenter systems and to grow them on an industrial scale. Plant cell cultures able to provide a source of secondary products would make the industry less reliant on an imported source of plants and enable it to regulate its own supply of raw materials more effectively. This economic drive provided the impetus to examine the productivity of large scale plant cell culture in detail.

12.1 Effect of culture conditions on secondary product synthesis

Cell suspension cultures and callus cultures, derived from plant species that normally accumulate a wide range of secondary products, showed that the concentration of the most important secondary products were all uniformly low. In addition, profiles of secondary products in tissue cultures were often unlike those in the mature plant. Subsequently, research was directed towards increasing yields of secondary products in callus and cell suspensions. The approach was the same as that used to boost yields in large scale microbial systems. It was based on a selection for high yielding cells and modifications to the medium and cultural conditions, including the conditions for the large-scale culture of plant cells. In time the research included cell immobilization, biotransformation, elicitation and genetic modification (*Figure 12.1*).

12.1.1 Composition of medium

In most batch cell suspension cultures, secondary product accumulation tends to increase at the end of the period of rapid cell

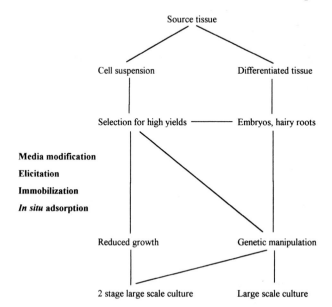

Figure 12.1. Sequence of stages in the stimulation of secondary product synthesis in plant cell cultures.

division of the growth cycle. In *Catheranthus roseus* cultures, the maximum increase in total indole alkaloids increased at the late exponential phase. The same was true for the tropane alkaloids in a number of Solanaceous species and rosmarinic acid in cell cultures of *Anchusa officinalis*. Even in cultures of species where accumulation of secondary products is growth related, inhibition of the growth of the culture leads to an increase in secondary product accumulation.

There appears to be an inverse correlation between growth and secondary product formation in both patterns of synthesis. It was necessary for cells to grow slowly for maximum secondary product synthesis to occur. The role of the media in achieving limited growth showed that the phosphate supply was critical. For example in *Catharanthus roseus* cultures, transferred to a medium without phosphate and 2,4-D, the indole alkaloids, ajmalcine and serpentine accumulated rapidly within 6–7 days of transfer. This accumulation was preceded by a large increase in activity of a key enzyme, L-tryptophan decarboxylase. Enhanced activity of the enzyme was required for the increase in indole alkaloid accumulation but substrate availability was thought to be as important. The effect of reducing the level of phosphate and 2,4-D was primarily to reduce growth and this seemed to be the trigger for the increase in secondary product synthesis [1].

Under conditions of limited growth, the concentration of secondary products shows an increase but the total amount may be small if the biomass is limited. In order to achieve maximum secondary product formation in culture, it would be necessary to accumulate a large initial biomass followed by a period in which growth is inhibited. This would require a two-stage culture process in which the initial medium is designed to stimulate growth and rapid accumulation of biomass, followed by a production medium that is nutrient-limited but that contains a high concentration of sucrose. The commercial production of shikonin, from cultures of *Lithospermum erythrorhizon* is used primarily in Japan in the cosmetics industry and for the treatment of burns. It is produced on a large scale from cell suspension cultures in a two-stage process with an initial growth stage and a final production stage.

12.1.2 Immobilization of cells

Where secondary product accumulation increased in cells showing limited or zero growth, cultures could be maintained in this state for a prolonged period as biosynthetic units. This is the principle of cell immobilization. Here the cells are enclosed in an inert material so that they are still in contact with each other and maintained in a medium that only allows a limited growth. Usually the medium has a reduced phosphate, nitrate and growth regulator supply and a large increase in the sucrose level. This medium is circulated around the immobilized cells so that secondary products released into the medium can be removed, thus reducing any feedback inhibition. Since the immobilized cells are growing very slowly, they can be maintained as an active production unit for up to 30 weeks compared with 3–4 weeks for a normal cell suspension culture (*Figure 12.2*).

The immobilization method may vary but all retain the principle of providing a protected environment for the cells.

1. Nylon sheets are folded to enclose the cells then the sheets are maintained in a bioreactor with an airlift system to circulate the nutrients.

2. Porous foam blocks are included in liquid medium containing a cell suspension. Cells grow into the spaces within the blocks, these blocks are then maintained as a bed while medium is circulated around them by an airlift system.

3. The cells are enclosed in a calcium alginate gel formed by adding drops of cell suspension and sodium alginate to a calcium alginate solution. Spheres of calcium alginate enclosing the cells can then be maintained in a fixed-bed column with medium trickling over the beads. The calcium alginate gel system has been used for the production of methyl digoxin from methyl digitoxin using cultures of *Digitalis purpurea*.

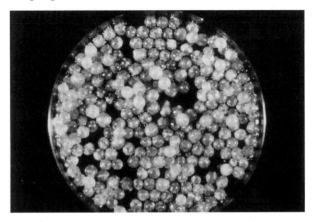

Figure 12.2. Cell suspension immobilized in Ca-alginate and maintained on a low-growth medium.

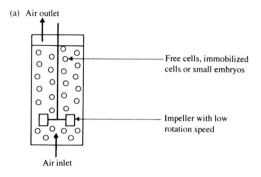

(a) Air outlet

Free cells, immobilized cells or small embryos

Impeller with low rotation speed

Air inlet

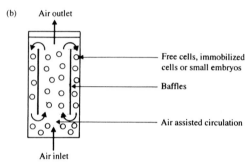

(b) Air outlet

Free cells, immobilized cells or small embryos

Baffles

Air assisted circulation

Air inlet

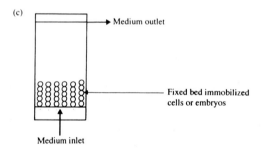

(c)

Medium outlet

Fixed bed immobilized cells or embryos

Medium inlet

Figure 12.3. Large-scale culture of free cells, immobilized cells and differentiated tissue.

Techniques have been developed to culture the immobilized cells on a large scale. Bioreactor designs have included the use of airlift fermenters which maintain the cell-filled beads or foam block beds, or flat-bed systems where the cells are trapped within membranes or fibers and the medium is run over the cells (*Figure 12.3c*).

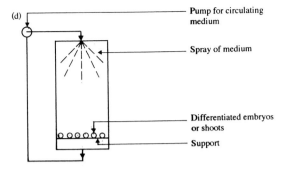

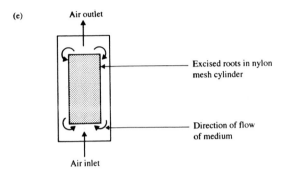

Figure 12.3. Continued.

12.1.3 Biotransformation of precursors

Because of the complexity of many of the secondary products, it has not been possible to synthesize them chemically. Simple intermediates have often been synthesized and converted to final compound *in vitro* with the aid of secondary product pathway enzymes extracted from plant tissue cultures and incubated with the precursors. Alternatively, the complete cells have been cultured in the presence of the intermediates or precursors. There are a number of examples [2].

1. Pathway enzymes leading to alkaloid production have been extracted from *Catheranthus roseus*, and incubated with chemically synthesized precursors to provide a method for improving the yield of the anticancer drugs, vincristine and vinblastine, which are found in small quantities in this plant.

2. These same enzymes are used as reagents in the commercial bioconversion of a substrate, dibenzylbutanolide, into the anticancer drug etoposide.

3. The parent plant, *Podophyllum peltatum* from which the etoposide is derived, has been used as a source of enzymes for the same biotransformation on a large-scale batch or semi-continuous cell culture process.

4. Two cardiac stimulants in the species *Digitalis pupurea* are digitoxin and derived from it enzymically is digoxin. Digitoxin is present in the greatest amount but is less effective as a heart stimulant. Digitoxin is extracted from the intact plant, is chemically methylated to methyl digitoxin to provide a substrate in an incubation medium of immobilized cell cultures of *Digitalis purpurea* which then biotransform the methyl digitoxin in the medium to methyl digoxin.

12.1.4 Differentiation of tissue cultures

In the transition from simple single cells and small cell aggregates in a cell suspension to larger aggregates, then differentiation into roots, shoots or embryos, the more complex tissue shows an increased capacity for secondary product accumulation. This improvement in secondary product accumulation seems to be associated with increased cell–cell contact and the increased complexity of cells and cell structures that are found in differentiated tissue. Thus, cell aggregates and suspensions of celery embryos are more productive of essential oils than undifferentiated cell suspensions (*Figure 12.4*), while differentiated shoots of *Digitalis purpurea* contain 1000 times more digitoxin than undifferentiated green cultures. Cultures of *Catheranthus roseus* showing shoot differentiation contain increased levels of serpentine and ajmalcine and even vincristine and vinblastine which are normally absent from undifferentiated cultures [1]. The effect of tissue and organ differentiation on secondary product formation is shown most clearly in transformed roots.

These structures are initiated through the action of a root pathogen, *Agrobacterium rhizogenese*. The bacteria enter the plant through wounds in the epidermis, then the bacterial plasmid, R1, transfers a portion of its DNA, the T-DNA into the plant cell and ultimately the T-DNA becomes incorporated into the plant genome. Expression of the host plant genome is modified by expression of the bacterial genes which cause overproduction of auxins and cytokinins. This leads to the production of a gall and a proliferation of roots. These roots can be excised and if placed on a simple inorganic medium will grow rapidly and continuously. The secondary product level is also comparable to that of the plant *in vivo* and much higher than in cell suspensions. Thus the yield of hyoscyamine is 80 mg l^{-1} in transformed roots

Figure 12.4. Differentiating cell cultures contain higher levels of essential oil than undifferentiated callus.

compared with 1 mg l^{-1} in cell suspensions. The procedure for initiating root cultures has now been modified so that large numbers of roots can be produced from cultured leaf material without the formation of a gall.

1. Leaf sections are removed from the parent species, surface-sterilized and are placed on a medium (MS medium, 3% sucrose) without growth regulators. The surface of the leaf is punctured with a needle, coated with a culture of *A. rhizogenese* and the leaves are allowed to remain on the medium. The *A. rhizogenese* must be of the appropriate strain in which the gene responsible for the synthesis of cytokinins (*tmr*) has been removed from the plasmid and a gene responsible for resistance to an antibiotic (kanamycin) inserted into the plasmid along with a promoter, CaMV35S (see Chapter 14).

2. When roots have grown from the surface of the leaf, these are excised and placed on a nutrient medium (MS, 3% sucrose) again without growth regulators but containing the antibiotic amphicilin (200 mg l^{-1}). The antibiotic in the medium removes the bacteria but the roots need to be subcultured a number of times on the antibiotic-containing medium before the bacteria are removed entirely.

3. Bacteria-free roots are grown in a liquid medium of the same composition but without the antibiotic. Growth of the roots is very rapid both in length and in the formation of multiple side shoots. The roots can also be grown on a large scale employing cylindrical bioreactors with the medium trickling over the roots. Growth is so vigorous that a dense mat is formed.

These transformed root systems have great potential for secondary product formation since they are amenable to selection and genetic modification. For example, on a medium containing auxin and cytokinins, the roots will initiate callus from which cell suspensions can be derived. These cell suspensions can be selected for high yields using the methods described earlier. Selected cultures showing an enhanced yield are then placed on a medium lacking growth regulators when the cultures will redifferentiate into roots. These roots will retain the character for enhanced yield, unlike cell suspensions which will tend to lose a selected character unless the culture is under constant selection pressure.

12.1.5 Elicitation of secondary product synthesis

Host plants respond to invasion by micro-organisms by the rapid synthesis of a number of secondary products and enzymes designed to restrict the spread of the invading pathogen. The trigger for the production of these secondary products (phytoalexins) and pathogenesis-related proteins is the presence of compounds called elicitors. These compounds are released from degraded cell walls of the pathogen, or host plant, and trigger a series of defense reactions in the host plant. The relationship between elicitors and host response has provided the biochemist and molecular biologist with a means to examine the complex world of pathogen attack and host defense.

It has also provided a mechanism by which secondary products may be stimulated in cultured cells [3]. The elicitors are not necessarily specific for the phytoalexins. For example the glucans, oligosacharides isolated from yeast, can act as elicitors to alkaloid production (berberine) in *Thalictrum rugosum* and sanguiarine in *Eschscholtzia californica*. The amount of elicitor added to the cell suspension is usually very small. The optimum concentration for berberine production was 200 µg elicitor (as carbohydrate) per g fresh weight of the cells. This had to be added at a specific stage of growth of the culture to achieve maximum response. The highest alkaloid production from both of these cultures was achieved when the elicitor was added at late exponential or an early stationary stage of the culture. This coincided with the period when the metabolic input into primary metabolism was being reduced. The response was very rapid and maximum levels of sanguinarine for example were observed within 6–8 h of elicitation of the *E. californica* cell cultures with an increase of sanguinarine of 8–10 times the unelicited cells. The response was only temporary and within a short period of a few hours, the alkaloid disappeared to be replaced by increasing amounts of another alkaloid, macarpine. Most of the increased sanguinarine alkaloid was released into the medium whereas the berberine was retained within the cell vacuole. This temporary burst of production is a common feature of elicitation.

In both the *T. rugosum* and *E. californica,* the temporary increase in alkaloid levels was matched by a similar increase in the enzyme tryptophan decarboxylase (TDC), which links primary and secondary metabolism. Under optimal conditions of elicitation, this enzyme was increased 10–20 times higher than in non-elicited cells. The increase was also sensitive to cycloheximide, indicating that it was regulated at the transcriptional level. In another member of the poppy family, *Papaver bracteatum*, the alkaloid sanguinarine was also responsive to a number of elicitor preparations. One elicitor preparation was a *Botrytis* culture homogenate added as 1 ml per 100 ml culture. Sanguinarine accumulation occurred after 12 h and after 79 h had accumulated sanguinarine equivalent to 3% dry weight with approximately 15% released into the medium. Once again this increase in sanguinarine appeared to be a result of induction of the control enzyme TDC.

The large increase in such an important secondary product prompted an investigation into the possibility of repeated elicitation of the cultures in successive culture periods. Cell viability did decrease after the third round of elicitation with a corresponding decrease in production. The decline in production of the alkaloids after the initial burst was thought to be due to feedback inhibition on the secondary pathway. Normally, many secondary products are stored in the vacuole or in specialized vesicles in the cytoplasm of the intact plant. In tissue cultures there is a partial release of these compounds into the medium where they may exert a feedback inhibition on the rate of secondary product synthesis within the cell.

The fact that a proportion of the accumulated product in unelecited and elicited cells was secreted into the medium has led to an investigation into the possibility of *in situ* adsorption or extraction of secondary products [4]. Water-insoluble products could be removed from the cells by the inclusion in the medium of inert hydrophobic chemicals (liquid or solid) which have a high adsorption capacity for hydrophobic plant products. A good example of an inert compound that can be added to the medium is *n*-hexadecane. This compound when added to a cell suspension culture of *L. erythrorhizon* increased shikonin production 8-fold with 95% of the shikonin being adsorbed by the *n*-hexadecane. When the cultures were elicitated in combination with *n*-hexadecane in the medium, the yield was increased 65-fold. Shikonin production was also increased 3-fold from hairy roots of *L. erythrrizon* by the same procedure of adding *n*-hexadecane to the medium. Growth of the hairy roots was better than the control cultures since removal of the hexadecane eliminated the growth inhibitory properties of the shikonin. Growth of the hairy roots on a relatively large scale (500 l) has been achieved in an air lift fermenter system where the roots have been contained in a permeable tube. The airlift fermenter system allows for easy addition of the elicitor and extractive compound at any stage of growth of the roots. Equally, removal of the spent medium containing the extracted compounds is not a problem.

12.2 Cell selection *in vitro* for high yielding cell lines

Yields per cell are determined by a complex interaction of factors. These are the cell genotype, cell age, position of a cell in an aggregate, the level of differentiation of the cells in a tissue and the composition of the media. In a cell culture the genotype of any one cell may differ from that of the parent cells as a result of somaclonal variation. Since somaclonal variation can affect the ability of a cell to synthesize secondary products, a culture may contain cells with a wide range of biosynthetic capacity. This variation provides the basis for selection for increased yields by a number of methods.

1. The direct visual method of identifying high yielding cells is simple and inexpensive. Cells that produce colored secondary products such as anthocyanin, betacyanin, or saffron are the easiest to screen for in this way since the most intensely colored cells can be identified directly. A cell suspension is plated onto the surface of agar containing the callus growth medium for the species. If the culture is present as a callus rather than as a cell suspension, then the callus must be broken into small fragments of 10–15 mg fresh weight on an agar plate. Cells, small aggregates or portions of an aggregate that show enhanced color formation are identified by examination of the petri dish under a binocular

microscope in a laminar flow cabinet, then the high yielding cells or parts of an aggregate transferred to new medium and the process of selection repeated. Colored secondary products are relatively rare so a more sophisticated screening procedure is required for most secondary products.

2. Because alkaloids are the most important group of secondary products, the direct visual method has been adapted to select for high-alkaloid-containing cells. This modification is based on the fact that alkaloids will fluoresce under UV light. A dilute suspension of cells is placed on the surface of nutrient medium in a petri dish as before, then the cells are examined with a microscope under UV light. Those cells or small aggregates which fluoresce most brightly can be easily identified then transferred to fresh media for a further cycle of growth and selection. This method of identifying high yielding cells has been automated using an automatic cell sorter so that large numbers of cells can be screened. For the automatic cell sorter to work correctly, the cell suspension must contain single cells and small cell aggregates since the suspension is passed through a narrow outlet creating a single stream of cells. High-yielding cells are identified by their enhanced fluorescence in a UV beam and are subsequently redirected into another container.

3. A less direct method which has been used very effectively is that of radioactive immunoassay (RAI). In this method the secondary product is attached to a protein then the protein conjugate injected into a rabbit where it acts as an antigen and generates antibodies specific for that secondary product. The separated antibodies are mixed with a solution containing the unknown concentration of secondary product and a known concentration of radio-labelled secondary product. Both radioactive and non-radioactive secondary products form a precipitate with the antibody, the radioactivity of which can be measured. The degree of dilution of the radioactivity from the activity of the original radioactive secondary product indicates the amount of non-radioactive secondary product present, that is, the amount of secondary product in the extract. Since the test is very sensitive, each small sample of cells due for analysis is split so that having identified high-yielding samples the remaining half can be subcultured and bulked.

Unfortunately, high yielding lines are not always stable so that on continued subculture and in the absence of a selection pressure for high yields, the average cell yield may decline. The reason for this is that the culture may revert back to the original dominant cell type which consists mainly of low-yielding cells. This rapid loss of high-yielding cells from the population does create problems when these cells are used as an inoculum in large-scale culture. As cells multiply in large-scale culture, the population of cells may change so that over the period of the growth of the culture, the average level of alkaloid per cell may decline and cause the yield to be less than that predicted.

12.3 Genetic modification of tissue cultures

The current methods of yield enhancement based on selection and modification to media and culture conditions may have reached a ceiling, so that any further increase in yields requires a more specific approach. In plant cells, this has involved identifying enzymes that link the secondary and primary pathways on the assumption that it is the activity of these key enzymes that control the flow of intermediates into the secondary pathways. An increase in the constitutive activity of these key enzymes could lead to increased synthesis of secondary products for the whole culture period and not just when growth slows down. It also assumes that the activity of the whole pathway is controled by one enzyme, which may not be true for the complex secondary product pathways in plants. It is possible to change the activity of one enzyme by introducing multiple copies of the gene for that enzyme. This will cause a constitutive increase in production of the enzyme and therefore enzyme activity (see Chapter 14) [5].

The best known key enzymes in plant cells are phenylammonia lyase (PAL), which controls entry into the phenyl propanoid pathway, and the amino acid decarboxylases such as tryptophan decarboxylase (TDC) which may control alkaloid synthesis. Harmalol is formed from tryptophan with tryptamine as an intermediate in the plant, *Peganum harmala*. In cell cultures of *P. harmala*, serotonin rather than harmalol, is found, whereas in differentiated cells the culture was able to form harmalol. The enzyme TDC was identified as the rate-limiting step in the synthesis. It was concluded that harmalol production could be increased if the enzyme was expressed constituitively rather than only on initiation of differentiation. The gene for tryptophan decarboxylase, (*tdc*), has been identified and isolated from another alkaloid producing species, *C. roseus*. The gene was cloned and the vector transferred to the root pathogen, *A. rhizogenese*. Seedlings of *P. harmala* were infected with *A. rhizogenese* during which process the *tdc* gene was transferred into the seedlings. Cell and hairy root cultures were derived from the seedlings and the tissues analyzed for

TDC activity. Analysis showed an increase in constitutive expression of the enzyme of 25 times that in the untransformed cells and an increase of 4 times in the roots. Although activity of the enzyme had increased, the level of the alkaloid accumulation was unchanged. It was suggested that there was a further rate-limiting step closer to harmalol synthesis.

In another approach [6], an attempt was made to increase the level of scopolamine in *Atropa belladonna*. Hyoscyamine and scopolamine are tropane alkaloids found in this species of the Solanaceae. Since hyoscyamine has an undesirable effect on the central nervous system, scopalamine is the preferred alkaloid and efforts have been made to raise the yield of this alkaloid. In plants, scopalamine is synthesized from hyoscyamine via the intermediate, 6 β-hydroxyhyoscyamine. The enzyme hyoscyamine 6 β-hydroxylase (H6H) catalyzes the reaction and is probably a rate-limiting step since there is a correlation between the activity of this enzyme and the ratio of scopolamine and hyoscyamine. The gene for H6H was inserted into *A. rhizogenese* which was then used to infect leaf explants to produce hairy roots. Examination showed that hairy roots which contained the inserted gene had 5 times the level of the H6H enzyme compared with the untransformed hairy roots. Alkaloid analysis demonstrated that the levels of scopalamine had increased 2–5 fold. Regeneration of roots into plants showed that the leaves of the transformed plants also contained higher levels of scopalamine than the untransformed plants.

The capacity to transfer this enhanced ability to other species was shown by transforming tobacco plants using the same route. Tobacco does not produce tropane alkaloids and neither do the plants, yet it expressed the H6H gene and the plants isolated from the transformed hairy roots also contained the H6H enzyme. The enzyme was active since both hyoscamine or 6 β-hydroxyhyoscyamine fed to the roots of transformed plants were converted to scopalamine. This showed that the ability to convert H6H or hyoscyamine could be transferred to any plant which can be easiliy cultured as cells or hairy roots. The transformed cells or roots could then be used to achieve a biotransformation of easily available precursors.

12.4 Large-scale culture of plant cells

The technology for the large-scale production of cell suspensions was based on the assumption that plant cell suspensions could be grown on the scale of industrial microbial systems. There are however a number of fundamental differences in the structure of plant cells which makes large scale culture of these cells technically much more difficult. For example, plant cells in culture are up to 200 μ which is 20 times larger than microbial cells, cultured cells exist as aggregates rather than single cells and have a doubling time of 2–5 days instead of hours. A large inoculum density of cells is required (5–20% by volume) but in comparison with microbial systems, the cell cultures have only a small oxygen demand. The consequence of these differences is that the plant cells are very sensitive to shear forces, show a rapid sedimentation, require large inoculation volumes, and because of the low growth rate have long inoculation runs with all the attendent problems of maintaining sterility.

In order to increase productivity, both yield and biomass accumulation need to be high. The practical maximum biomass is 40–60 g l^{-1} and there should be a minimum yield of at least 2% of the secondary product as total dry weight of the cells for the system to become economic. The yields of many cell cultures are known to exceed this figure although not always when grown on a large scale. Thus high yields have been obtained for rosmarinic acid (*Coleus blumeii*) 21.4%, anthraquinone (*Morinda citrifolia*) 18%, benzylisoquinolines (*Coptis japonica*) 15%, shikonin (*Lithospermum erythrorhizon*) 12%, berberine (*Berberis wilsonae*) 10%, shikimic acid (*Gallium mollugo*) 10%, diogenin (*Dioscorea deltoidea*) 8%, nicotine (*Nicotiana tabacum*) 5% [7].

Because of the difficulties with shear and sedimentation, the bioreactor design has favored an airlift operation. Here a stream of sterile air is forced up through the cell suspension thereby mixing the cells and maintaining them in a state of suspension (*Figure 12.3a*). With this reactor design, it was easier to maintain asepsis and also cheaper to run. However, the airlift system did not work well with very dense suspensions where pockets of unmixed cells formed. Other, more conventional, stirred reactors have been investigated and are now looked upon more favourably after it was found that some plant cell cultures were resistant to the high shear stresses generated in stirred-tank bioreactors (*Figure 12.3b*). Thus a stirred-tank bioreactor of 5000 l has been used to maintain cultures of *Catheranthus roseus*. Partially differentiated tissue grown *in vitro* was seen as an alternative system to large-scale culture of undifferentiated cell suspensions for the accumulation of secondary products. This is because in differentiated cultures, the synthesis of the normal profile of secondary product is triggered and yields are comparable to intact plants. The cultures have acceptable yields and at an earlier stage of differentiation can be grown *in vitro*, although the large scale culture of differentiated embryos, shoots or roots does present a problem. Where these are small units such as early stage embryos, it has been possible to use modified bioreactors (*Figure 12.3c*). Stirred tanks with large slowly rotating vanes or rotating horizontal drums with a series of internal baffles have also been successful. Where the differentiating culture is producing shoots, the culture is static and the medium is added from above, then the spent medium is collected below the culture and recycled (*Figure 12.3d*).

Transformed hairy roots are also shear-sensitive and a suitable design is to provide a protective tube of polyurethane foam or nylon mesh in which the roots can grow. The tube is then maintained in an airlift reactor which aerates and mixes the medium around the central tube (*Figure 12.3e*). Any recycled medium is very accessible and can be replaced easily with fresh medium and the old medium extracted for any released products.

The technical problems for initiating and maintaining cell suspensions in large scale culture are gradually being overcome. The alternative strategies of cell immobilization and elicitation. for enhancing yields are also successful and can also be carried out in large-scale culture. However, the technique still requires a great deal of development before these methods become a commercial reality. One of the techniques that is nearest to commercial exploitation is the combination of chemistry and biotechnology where the cells perform the metabolic changes in the intermediates that the chemists cannot. In time the complete secondary pathways will be known, and it may be feasible to modify a number of the enzymes by genetic engineering techniques so as to enhance the activity of the complete secondary pathway and thereby increase yields very significantly.

References

1. **Collin, H.A.** (1987) In *Advances in Botanical Research,* Vol. 13 (ed J. Callow). Academic Press, London, pp. 145–87.
2. **Kutney, J.P.** (1995) In *Current Issues in Plant Molecular and Cellular Biology* (ed. M. Terzi, R. Cella and A. Falavigna). Kluwer Academic Publishers, Dordrecht, pp. 611–616.
3. **Nishi, A.** (1994) In *Advances in Plant Biotechnology,* Vol. 4 (ed. D.D.Y. Ryu). Elsevier Science, Amsterdam, The Netherlands, pp. 307–338.
4. **Chang, H.N. and Sim, S.J.** (1994) *Ibid,* pp. 355–369.

5. **Berlin, J., Fecker, L., Herminghaus, S. and Rügenhagen, C.** (1994) *Ibid*, pp. 57–81.
6. **Yamada, Y., Yun, D.J. and Hashimoto, T.** (1994) *Ibid*, pp. 83–93.
7. **Scragg, A.H.** (1993) In *In vitro Cultivation of Plant Cells, Biotechnology by Open Learning*. Butterworth-Heinemann, Oxford, pp. 151–178.

13 Micropropagation techniques in horticulture and crop improvement

Vegetative propagation is of considerable importance to agriculture, horticulture and forestry since it provides for the production of uniform material for crop planting, the multiplication of elite food trees and ornamentals, vegetables and soft fruit stock, and the propagation of forest trees. Vegetative propagation is the normal method of reproduction for the flowering bulbs, which commercial practise then exploits. In contrast in woody perennials, vegetative propagation is used to overcome the limitations imposed by long breeding cycles. Methods of crossing, selection and seed production used to improve annual crops are unsuitable for perennials because selections can only be made when the perennial is mature which for trees may be a minimum of 40 years. Instead, selected elite individuals are identified and multiplied vegetatively.

These propagation techniques do produce the occasional rogue individual which is normally rejected, or which may be used if it is of novelty value. This variation is a result of somatic mutations originating in the apex and may appear as different leaf shapes, or flower shape and color, or fruit shape, size and color.

Propagation is by the traditional practise of cuttings using shoot tips and nodal axillary buds as source material, or from the production of adventitious shoots from cut stems or leaves. Bud cuttings are produced by excising the terminal shoot tip and segmenting the remaining shoot into nodal portions so that each will contain an axillary bud. The basal part of the shoot is then treated with a rooting hormone and the treated portion inserted in soil and placed under a cover to reduce transpiration but in sufficient light intensity to enable photosynthesis to proceed. The bud expands and grows into a new shoot and the basal part of the root forms new roots. Adventitious shoots arise as a result of an internal re-differentiation within already differentiated stem or leaf tissue in reponse to cutting and when large enough, the shoots are excised and rooted as normal cuttings.

The criticism of the traditional method of macropropagation is that it is labor-intensive, of limited productivity and to some extent is controled by the seasons. In an effort to increase productivity, alternative micropropagation methods have been developed. These involve the use of tissue culture, and are based on the same traditional model of shoot tip culture, axillary bud culture, adventitious shoot production, and a new approach based on the use of embryo culture in liquid medium (*Figure 13.1*).

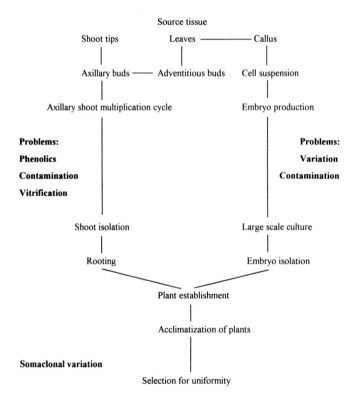

Figure 13.1. Sequence of stages in the production of plants by micropropagation.

13.1 Source tissue for micropropagation

It is important to use healthy, vigorously growing plants as source material. In some instances, however, the state of dormancy of the bud and the age of the parent plant will have more of a pronounced effect on the response of the source tissue to the medium than the health of the tissue.

13.1.1 Effect of bud dormancy

All temperate woody perennials show a seasonal cycle of bud activity. The terminal buds are active in the spring, producing leaves and the axillary buds will also produce leaves unless the plant is strongly apically dominant. Late summer is characterized by a decline in meristem activity of the buds as they become dormant in response to internal changes. This stage of dormancy is temporary and is released easiliy by excision of the buds and transfer to conditions *in vitro* (e.g. 12 h dark and a constant temperature of 25°C). Later in the season, the dormancy of buds on the intact plant becomes permanent in response to the seasonal decline in daylength. The buds must then be exposed to a prolonged period (8 weeks) of low temperatures (less than 10°C) before the dormancy is released. For the best response to a micropropagation procedure, it is advisable to use tissue and buds from non-dormant plants.

Many tropical trees and some temperate trees, such as the oak, show a flush cycle whereby the buds will show regular periods of active growth and dormancy within the year. This cycle of bud growth and dormancy seems to be regulated by internal mechanisms since it will occur in a constant glasshouse environment. The response of excised nodal and terminal buds to the composition of a nutrient medium will depend on the stage of flushing of the buds. In experiments to identify the correct media composition for growth of shoot tips or nodal cultures it is important to use buds that are all at the same stage of flushing. The period when the bud is emerging from the temporary period of dormancy is likely to be the period of maximum response.

13.1.2 Effect of juvenility and maturity

The stage of maturity of woody plants also affects the response of the tissue, or bud, to a tissue culture medium. Thus, tissue cultures from seedlings and young trees are more likely to regenerate from callus than if the explants are taken from older trees. Shoot tips and nodal segments will also form axillary shoots and roots more readily where seedling or young trees are the source material. This difference in response is caused by physiological differences between the younger juvenile and older mature plants and it has serious consequences for forest tree breeding. Selection of elite characteristics in timber trees occurs when the trees are mature and well after the juvenile stage, which makes it difficult to propagate from these individuals.

In order to overcome this problem, the state of maturity of the older tissue is reversed by severe pruning and also by repeated grafting of buds from mature scion wood onto seedling rootstocks. Regrowth from pruned shoots and from grafted buds is more juvenile in character and can be used as an explant source for callus cultures and for shoot tips and nodal segments. Repeated subculture of axillary shoots derived from mature tissue also improves the rooting ability of the axillary shoots.

An alternative approach is to establish molecular markers for these complex and polygenic characters in mature individuals, such as disease- and pest-resistance, height, timber quality, rate of growth, and resistance to environmental stress. Subsequently juvenile progeny are examined for these markers, then juvenile plants, which possess markers specific for a desired character, are easily propagated by standard or micropropagation methods. The molecular markers make use of band patterns from restriction fragment length polymorphism (RFLP), RAPDs and micro-satellite analysis that correlates with a particular character. The added advantage for perennials is that since selection take place at the seedling or juvenile stage rather than at plant maturity which may be many years later there is a great saving in time.

13.2 Methods of micropropagation

Methods used in micropropagation are based on the production of sterile tissue, stimulation of regeneration, rapid growth of the young plantlets, rooting of the plantlets and their weaning onto normal soil conditions. The techniques make use of methods of sterilization of explants and of regeneration of callus described in earlier chapters (Chapters 3–5). Further details of the methods are given below.

13.2.1 Shoot tip and axillary bud culture

This micropropagation method is a scaling down of the traditional *in vivo* approach to the production of clonal plants and is the one preferred for commercial propagation because of its simplicity. The method is as follows:

1. Parent plants should be growing vigorously with no signs of bud dormancy, be disease- and pest-free and should not have experienced prolonged water or nutrient stress.

2. Shoot tip or nodal sections with lengths not exceeding 10 mm are excised then surface-sterilized using the methods described in Chapter 5. Because the plant has been growing in a non-sterile environment, there may be a number of explants lost through insufficient or excessive surface sterilization. It may be necessary to take an empirical approach to establish the optimal concentration of sterilant and duration of exposure. Removal of the outer layers of primordial leaves of the shoot tip prior to sterilization may reduce the potential sites of infection. Usually the axillary buds are too small to attempt to remove the outer leaves.

3. Stem tip and nodal section after surface sterilization are implanted base down into medium (MS medium, 0.05–0.5 mg benzylaminopurine, 1% sucrose, 8 g l^{-1} agar) in a Universal vial or plastic container. The medium contains a high concentration of cytokinin to overcome the apical dominance of the bud. Any plant which normally produces axillary shoots will produce 5–10 shoots every 4–8 weeks under these conditions.

4. When the axillary shoots are of a suitable size (10–20 mm long) each is excised and transferred to fresh medium of the same composition and a further cycle of shoot multiplication begins. This cycle can continue almost indefinitely, but there may be a loss in vigor over time in which case new isolates should be used.

5. Shoots are removed from the cycle when they are 20 mm long and are stimulated to form roots by dusting the bases of the excised shoots with a commercial rooting powder (indole-3-butyric acid (IBA) in talc powder) then the shoots planted in a conventional rooting compost as normal cuttings.

6. Alternatively root production is initiated *in vitro*. Excised shoots are transferred *in vitro* to a medium in which the high cytokinin has been replaced by high auxin (naphthaleneacetic acid (NAA) or

indole-3-butyric 0.5–10.0 mg l^{-1}) to rapidly initiate root formation, then the shoots are transferred *in vitro* to a medium with no growth regulators.

7. Growth of the excised shoots, or shoots rooted *in vitro*, may be checked at this stage as new roots are formed. If the individual shoot is planted in a hollow cellulose rod and is incubated in a covered container in an inorganic liquid media such as 0.1 M MS basal salts, this check is reduced. The container is maintained in a temperature- and light-controled environment to reduce the possibility of excessive transpiration. Since the above rods are biodegradable, the plantlets can be easily transferred to soil, whilst remaining in the rods.

8. When 100 mm long, all rooted plants are suitable for sale directly, or as a source of conventional cuttings.

Because of the higher cost of production compared with traditional methods, micropropagation is only used on the more expensive plants. These are, in order of importance, ornamental shrubs, followed by house plants then outdoor herbaceous plants. The criteria used to decide the appropriateness of the technique for any plant are the cost of production and the market value of the final plant. Increasingly however, this technique is being used to propagate timber trees such as larch, and food trees such as apple.

13.2.2 Meristem culture

Plants often carry endogenous infections of bacteria, fungi and viruses which must not be transmitted to stock plants or clonal products. The extreme tip of the plant, the meristem, is free from these infections. If the apical dome (0.1–0.5 mm high) is dissected out then treated in the same way as a shoot tip culture, the plants derived from it will be largely contamination-free. This technique is used for the commercial production of virus-free strawberry stock plants and seed potatoes.

In the latter crop, potato shoot tips provide a source of axillary shoots which, when rooted, can be induced to form microtubers (30–50 microtubers per explant in 4 months). Microtubers only weigh 10–30 g which means that they can be distributed commercially at a very low cost and since they can be stored dry, production can continue throughout the year. The microtubers are virus-free and the field performance of micropropagated tubers is comparable to plants grown from conventional disease-free tubers.

13.2.3 Adventitious bud production

Adventitious bud production occurs where non-meristematic cells become meristematic in response to growth regulators in the external medium, and nutrient gradients within the tissue. This combination of influences encourages a localized redifferentiation of cells (meristemoids) in the surface of the tissue. These meristemoids then develop as stem or roots depending on the ratio of auxin and cytokinin in the external medium. The production of meristemoids represents the initial stage of organ or even embryo formation and occurs in callus and in intact plant tissue. Details are as follows:

1. Shoots and embryos that form readily on callus of some species in culture (Chapters 6 and 7), notably tobacco in the Solanaceae and carrot in the Umbelliferae are adventitious in origin. Histological evidence would suggest that the individual shoots and embryos are of single-cell origin.

2. Embryo production occurs in some species from cells and cell aggregates in liquid culture, each embryo possibly derived from a single cell. This method of micropropagation is potentially very productive since the method is amenable to large-scale production. The cultures are maintained in large fermenters where the development can be synchronized by changing the composition of the media and cultural conditions. At a specific stage of development when the embryos require a less complex medium, that is, early torpedo stage, the embryos are transferred in small numbers to a development medium based on MS inorganic salts with all the growth regulators omitted. The embryos then develop into plants as normal. This system is being developed for coffee.

3. Adventitious buds are produced directly from parent leaf and stem sections when explant tissue is placed on a nutrient medium with the correct combination of auxin and kinetin. Shoots or roots are produced directly from cells within the tissue. Shoots are formed when the nutrient medium is high in cytokinin and low in auxin. The nutrient medium required to stimulate adventitious bud formation can be established by carrying out a Latin Square analysis where both auxin and cytokinin are varied (Chapter 5).

A well-known example of adventitious bud formation on intact plant tissue is African Violet which can produce very large numbers of adventitious shoots on leaf explants (Chapter 7). Large numbers of Begonia and Gloxinia are also produced by the same technique.

In the ornamental bulbs (Narcissus, Crocus Freesia, Lilium), micropropagation is by the production of adventitious buds on explant tissue. This may be bulb scales, stems or buds. Numerous adventitious buds are formed at the base of the scale and on the cut edges of the stems, or from the surface of the buds. Buds give rise to vigorously growing plantlets each of which forms a bulblet or cormlet.

13.3 Problems that arise *in vitro* during micropropagation

Because micropropagation is a largely commercial enterprise, there is pressure to produce a large numbers of uniform plantlets rapidly. This aim has focused attention on the problems that arise in tissue culture practise and the methods that are used to overcome them [1].

13.3.1 Removal of phenolic compounds

Phenolic compounds are produced by the plant in response to stress. Many of these compounds are phytotoxic and will lead to death of plant tissue if released onto cells. Any treatment that encourages phenolic production is undesirable. Thus, plants under water- or nutrient-stress, or diseased plants will have an enhanced phenolic level. Oxidation of the polyphenols by polyphenol oxidase when the tissue is excised causes increased levels of browning in shoot tip and nodal isolates which in turn will damage the plant tissue locally and inhibit the response to the medium. There have been a number of attempts to reduce the effect of phenolics whose presence is a particularly serious problem in the culture of woody perennials.

1. Source plants must be healthy and disease- and pest-free, be well-watered and have an adequate nutrient supply for at least two weeks before the plants are excised.

2. Phenolics are either adsorbed or leached out of the explant. Charcoal or polyvinylpyrrolidone added to the medium adsorbs the phenolic compounds and so reduces tissue blackening. Charcoal however has the disadvantage that it also adsorbs growth regulators. Alternatively the phenolics are leached into the medium by washing the excised tissue in running water for 2–3 h, or incubating the tissue in sterile water overnight, or repeated transfer to fresh medium every 2–3 days. The latter is particularly effective if liquid media are used for these initial

transfers. Reducing agents added to the medium (ascorbate, citrate, dithiothreitol and glutathione) prevent blackening by polymerization of phenolic quinones, thereby removing one of the substrates that leads to blackening of the tissues.

3. The activity of polyphenol oxidase is inhibited by a number of strategies. The inclusion of a chelating agent such as ethylene diamine tetra-acetic acid (EDTA) is able to chelate copper in the medium. This metal is required by the polyphenoloxidase as a cofactor but the period of inclusion must be limited to a week since copper is a trace element essential for the growth of the culture. Various other strategies lower polyphenol oxidase activity such as reducing the potassium nitrate levels or all of the inorganic salts by 50%, the use of reduced temperatures (e.g. 20°C) and the maintenance of the newly initiated cultures in darkness for one week. Once the the excess phenolic compounds are remetabolized they are not available for oxidation so that the browning will be less and after a period in a modified medium or culture conditions, the culture can be placed on a normal medium under standard conditions to optimize growth.

13.3.2 Detection of endogenous contaminants

Most of the problems caused by contamination of the newly initiated cultures derived from intact plant tissue are from endogenous viral, bacterial and fungal contaminants rather than a poor aseptic technique. The evidence that endogenous contaminants are present is that the infection only becomes apparent at a late stage in the initiation of the culture or after a series of subcultures, or the tissue continues to look brown despite changes to the media composition. Where there is an endogenous contaminant it is likely to be a result of a number of different organisms. It is important to check a sample of all the isolates so that the problem can be detected early in the isolation procedure. The approach used is as follows:

1. The shoot tips, or nodal segments, or explants are isolated using a standard surface sterilization procedure and a small sample is transferred to a variety of test media for two weeks. The remaining explants are transferred to standard growth media to produce axillary shoots, or to initiate callus depending on the path of micropropagation being followed.

2. The basal ends of the test sample of shoot tip, nodal segment, or stem explant are implanted in a range of media (nutrient glucose

media, potato dextrose agar, peptone yeast agar) to stimulate bacterial growth, and MS basal salts plus 1% sucrose to stimulate fungal growth. The presence of bacteria is confirmed by their colony appearance as well as their response to Gram staining, and fungal contaminants by their colony appearance and microscopic structure.

3. Where the contamination is detected on the test media and not on the normal tissue culture media, then the organisms are likely to be slow-growing endogenous contaminants.

4. If the test sample of shoot tips and nodal segments is free of contaminants, a sample of first generation of axillary shoots is transferred to the above bacteriological and fungal media for two weeks and samples of isolated callus to the same media.

5. The second generation of axillary shoots are produced and the callus subcultured and a sample inoculated onto the test media.

6. If test samples are still free of contamination then it can be assumed that there are no serious endogenous contaminants. However it cannot be assumed that the culture is entirely free of contaminants.

13.3.3 Removal of endogenous contaminants

There are various methods by which endogenous contamination can be removed. It is very important to do so, otherwise stock plants become contaminated and the contaminant is transferred to all clonal individuals isolated from them. Contamination can lead to poor growth of the propagated plants which makes them unattractive and risks the possibility of cross-infection to other species.

1. All sources of contamination (bacterial, fungal and viral) can be eliminated by isolating meristems from the parent plant as described earlier. The stock plants then form the source of plants for shoot tip culture or macropropagation.

2. The bacteria are removed from the culture by the use of antibiotics added to the medium, such as rifampicin (50–100 mg l^{-1}) but preferably a mixture such as cephotaxime, tetracycline and Polymixin B, although the latter may lead to further browning. The efficiency of each in combating a group of contaminants would have to be established for the particular tissue isolate. Antibiotics do have an inhibitory effect on plant tissue culture growth and

may also be mutagenic so their use must be seen as a temporary addition to the medium.

3. Fungal contamination can be reduced by the incorporation of benomyl. Viruses may be treated by the inclusion of antiviral agents such as Virazole syn. and Ribavirin and also by heat treatment of the bud cultures (30–40°C for 2–10 weeks). The exact details of the treatment are available in the literature or must be established by experimentation.

13.3.4 *Vitrification of plantlets* in vitro

The conditions within the culture vessel of micropropagated plants are extreme since the plants are in an enclosed atmosphere on a medium that, as a result of the sucrose and macro salts, has a high osmotic pressure. The consequence of enclosure is that the level of CO_2 is much higher than in the atmosphere (14% recorded) and there are very high levels of ethylene (2–3 ppm) and water vapor. The high ethylene levels will lead to morphological abnormalities such as stem thickening and altered orientation of the stem, root and leaves with respect to gravity. The high humidities may well discourage wax formation on the surface of the leaves and high osmotic pressures of the medium will lead to succulence development so that the tissue will appear thickened and translucent. These mishapen and thickenened leaves and stems have been referred to collectively as vitrification. In this form the shoots and plantlets do not transfer to soil very easily.

The simplest method of control, or rather amelioration of vitrification is to reduce the temperature of the base of the container to encourage condensation of water vapor, thereby creating a gradient of water retention in the agar. Other methods seem to reduce or remove the ammonium from the medium.

13.4 Problems that arise *in vivo* after micropropagation

Rooting of *in vitro* grown axillary shoots and advetitious shoots is a major stage in micropropagation and often it is stimulated under *in vivo* conditions similar to the methods used to stimulate roots on cuttings. Commercial production insists that roots are formed on the shoots rapidly then the plant transplanted to soil with the minimum check in growth and the minimum loss of individuals. In a commercial micropropagation procedure, it is also important to produce large numbers of identical clonal products that are also true-to-type to the parent stock. Standard vegetative propagation can achieve this level of uniformity so any micropropagation method must be comparable if it is to be acceptable.

13.4.1 Rooting of plantlets

There is a great deal of variability between species in the rooting response. Axillary shoots of the potato root very readily, but axillary shoots from the woody perennials do not. The major problem is when no roots are formed at the base of the shoot. This is most likely due to a lack of root initiation as a result of an imbalance of growth regulators. The critical growth regulator is auxin which is normally applied as a rooting powder to the base of the shoots *in vivo*, or as a brief immersion in an auxin solution *in vivo* (e.g. 0.01 mM IBA for 4 days) or as a component of the medium *in vitro*. Excess auxin *in vitro* will result in the production of callus rather than roots. Excessive callus production prevents direct vascular connections being made between shoots and roots so that after transplanting to soil the plant will die as a result of transpiration loss. Where callus production has occurred, it can be removed surgically prior to the shoot being subcultured into a more suitable rooting medium.

Roots can be stimulated to grow *in vitro* but it is uncertain how effective they are *in vivo*. The roots produced *in vitro* lack root hairs and it is difficult to remove agar from them without damage. Following transfer to the soil, the roots are thought to die and new functional roots develop. Both the presence of the agar and the loss of the original roots may increase the risk of infection entering the plant. On balance it is probably safer to use *in vivo* methods for stimulating root production. Transfer of the shoot to *in vivo* conditions is into standard potting compost and once rooting has occurred, growth is stimulated by the addition of liquid fertilizer.

13.4.2 Acclimatization of plantlets

The greatest danger during the period of root induction is the potential water loss from the shoot when the plant has no effective root system. Transpiration loss must be reduced as much as possible by maintaining a high humidity in the atmosphere around the transplanted shoots, then the plantlets must be acclimatized to the lower humidity conditions outside of the tissue culture container. Transpiration is reduced by creating a high-humidity fog around the plants. Water under pressure is forced out of a series of fine nozzles on overhead lines which produces a fog of droplets (10–30 μm diameter) that remain in suspension in the air for a period of time depending on the size of the droplet. The fog is released automatically either on a time basis or in response to a humidistat. For health and safety reasons the water must come directly from the mains.

Once roots have been established, the plants must then be acclimatized. This will involve a progressive reduction in the time exposed to the fog and an increase in the light intensity. Ambient temperature and base heat (i.e. soil temperature) are other variables which determine the success of the acclimatization process. Details of these can be established from the literature as they will vary with the species.

13.5 Variation in clonal plants

In a commercial micropropagation procedure it is important to produce large numbers of true-to-type identical clonal plants. Variation in micropropagated lines can arise from a number of sources (see Chapter 11).

1. Somatic mutations already present in individual cells of the original parent are passed on to callus cultures initiated from this explant tissue. Embryos and shoots derived from single cells in the callus will consist of cells carrying these mutations and any changes will eventually be expressed by the regenerated plants.

2. Initiation of callus and cell suspensions appears to generate somatic mutations in the cultured cells which are then transmitted to any regenerated embryos and shoots and subsequently to plants.

3. Adventitious buds arise from single cells or a small group of related cells in the parent. If these cells contain somatic mutations, these will be passed directly to the meristem of the adventitious bud and will thus affect the shoot and plant.

4. Conditions of culture in the early stages of micropropagation may modify the expression of the genome. These epigenetic changes may affect quite drastically the morphology of the regenerated plants but the changes disappear in the progeny or in perennials after a number of years of vegetative growth.

5. Somatic mutations may occur in cells of existing meristems or buds which are passed directly to the axillary shoots derived from these modified shoot tips or axillary buds. This has been the origin of sports.

The most stable system is the shoot tip and nodal bud culture. In this technique there is no loss of differentiated tissue or basic plant structure. The proportion of abnormalities, or sports, generated by this method of micropropagation is similar to the standard method where abnormalities are very rare. This is one of the reasons why the shoot tip method is still the preferred one for the commercial production of large numbers of identical individuals.

Where the propagated tissue is cultured as undifferentiated tissue for any period of time, there is a possibility that somaclonal variation will be generated. Such was the case with the oil palm. In this procedure, elite individuals are multiplied by a micropropagation program in which the *in vitro* stage is an embryogenic culture. Abnormal-looking plantlets are discarded and only normal-looking plantlets are allowed to grow to a larger size. Unfortunately, normal-looking young plants developed abnormalities as they matured. Analysis of the behavior of the mature individuals has shown that some of the unusual leaf and plant shapes and fruit structure do revert to a normal pattern whereas other changes are permanent. Using genome analysis at an early age, it should be possible to identify in perennials, such as the oil palm, the potential for developing abnormalities at a later period of growth and to discard these individuals.

In the oil palm culture, differentiating meristems may screen out the most extreme abnormalities. At present the use of embryogenic cell suspension cultures as a source of clonal individuals is being discussed in view of its higher level of productivity and potential for automation. Cell suspensions are less stable than callus tissue so there might be a greater problem of somaclonal variation in this method of multiplication. However, providing there is effective screening for identifying potential abnormalities, there should be no problem with a liquid culture procedure.

13.6 Advantages and disadvantages of micropropagation

Where a new method such as micropropagation is replacing an existing and successful propagation procedure within the industry, there must be distinct and clear advantages [2]. These are:

1. Rapid multiplication rate of easily regenerated species. Shoot tip culture can provide 5–10 axillary shoots every 4 weeks for each original bud. The cycle of axillary bud excision and shoot multiplication soon provides very large numbers of plants.

2. The lack of seasonal restrictions on plant production since the propagation stage is is laboratory-based. This applies particularly to houseplants which may be bought at any time of the year. Other plants which require outdoor planting are still seasonally determined.

3. Production of difficult-to-propagate plants so that new varieties and species can be introduced to the market.

4. Benefit to seed-propagated vegetables through micropropagation of elite individuals selected in the field. Thus, well-shaped and blemish-free vegetables such as cauliflowers can be selected under field conditions then multiplied by micropropagation to provide a sufficient number of plants to generate seeds.

5. Maintenance of self-incompatible inbred lines used in hybrid seed production. Micropropagation provides a method of maintaining the lines without the need for manual pollination such as for maintenance of inbred lines of cabbage and Brussels sprouts. Plantlets would only be grown to maturity when a flowering stage was required for hybridization and seed production.

6. The production of virus-free stocks such as rhubarb, strawberry and potato by meristem culture. This technique can provide stock plants in the case of strawberry which can then be multiplied by conventional cuttings. Yield increases of healthy, disease-free plants are very significant and the improved appearance of disease-free ornamentals an advantage.

7. The potential for automation through the use of adventitious bud culture and large-scale embryo culture in liquid medium. The latter may also involve the production of synthetic seeds in which single somatic embryos are surrounded by a coating of agar-containing nutrients. This compact agar layer provides an immediate nutrient and water supply to support the initial growth of the stem and root apex, thus allowing individual seeds to be sown in the same way as normal seeds.

The economics of initiating a micropropagation facility are determined by many factors amongst which must be a comparison with the cost of the standard procedures. Bearing this in mind, the most immediate choice of plants is going to be the expensive houseplants (Nephrolepis, Cymbridium, Gerbera, Saintpaulia), ornamental shrubs (Roses, Magnolia) and outdoor herbaceous plants (Primula, Pelargonium).

The disadvantages of the micropropagation system are:

1. Delicate product possessing a thin cuticle and few root hairs, which requires skillful handling by trained staff.

2. Expensive large-scale facilities required for media preparation, autoclaving, provision of aseptic conditions, temperature- and light-controled growth rooms and conditions for rooting and acclimatization.

3. Complex care of mother stock to prevent pest and pathogen contamination.

4. Prolonged period required after the *in vitro* phase to root and acclimatize the plants.

5. Extensive period of research and development to establish conditions of culture for new species or varieties that are difficult to regenerate or root.

The capital cost of setting up the facilities, providing staff and maintaining a research and development facility is going to be high.

A number of different types of business have emerged to meet the wide range of demands. These are based on the propagation for the nursery or horticultural business, the provision of specialist plants, the production of plants for seed production by multiplying up selected lines and the application of haploid breeding and genetic manipulation to crop improvement.

References

1. **Alderson, P.G. and Dullforce, W.M.** (1986) *Proceedings of the Institute of Horticulture Symposium, University of Nottingham, School of Agriculture.* University of Nottingham Trent Print Unit, Nottingham, UK.
2. **Giles, K.L.** (1991) In *Horticulture – New Technologies and Applications, Current Plant Science and Biotechnology in Agriculture* (eds J. Prakash and R.L.M. Pierik). Kluwer Academic Publishers, London, UK, pp. 155–160.

14 Genetic manipulation of crop plants

Plant breeders have been able to improve crop yields indirectly by introducing increased resistance to disease and environmental stress into susceptible varieties. This has been achieved by hybridizing susceptible varieties, either with more resistant varieties, or with a related wild species. Hybridization is followed by a programme of repeated backcrossing, or selfing, of the F1, combined with selection for the introduced trait. This process has been very successful, but it is time-consuming and expensive. In addition, plant hybridization may not always be possible because of incompatibility within and between species.

Various *in vitro* techniques have been developed to shorten the time required for backcrossing and selection and to overcome incompatibility problems. Protoplast fusion was seen as a method by which very widely separated species could be crossed without the problems caused by incompatibility reactions of a pollen fertilization. Unfortunately the early promise of this technique was not met. The genomes in fused protoplasts of widely separated species were often unstable and most, if not all, of one genome was often lost. The initiation of somaclonal variation by a period of tissue culture is a method for introducing new variation into a parent crop without the need for hybridization. Somaclonal variation is a significant source of variation in the regenerant generation, but the level of variation is much reduced in the progeny and it was not always stable in further progeny generations.

The other source of new variation using *in vitro* systems is that generated through plant transformation. Single genes are inserted into the genome of the parent crop species to change the resistance to disease, herbicides and environmental stress and also to change patterns of plant development, such as ripening. The genetic modification is directed and specific. There is no period of selection and the changes appear to be stable under field conditions. At the moment it is only possible to transform those characters that are controled by a single gene but in the future it is likely that linked polygenes will be transferred into crop species.

14.1 Strategy for plant transformation through gene transfer

While there is a wide range of single genes being used to transform crop plants, each transformation process shows a common strategy (*Figure 14.1*). This is as follows:

1. Selection of a character based on a single protein. For example, the mode of action of many herbicides is to inhibit the action of a single enzyme in the plant. Glyphosate for example inhibits the action of the enzyme, 5-enolpyruvyl shikimate-3-phosphate synthase (EPSPS), which controls the production of the amino acids phenylalanine, tyrosine and tryptophan.

2. Modification to a single gene. To increase crop resistance to glyphosate it is neccesary for the plant to overproduce the EPSPS enzyme, to express a modified EPSPS enzyme, or for the plant to

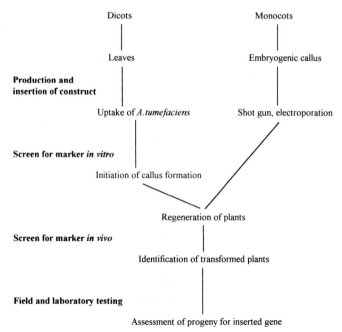

Figure 14.1 Sequence of stages in the genetic transformation of dicot and monocot plants.

show increased degradation of the herbicide. In the case of glyphosate, the most effective strategy for enhancing herbicide resistance was for the plant to express a modified EPSPS enzyme that would not be inhibited by the herbicide.

3. Identification of a single gene. In the example of glyphosate resistance an altered enzyme was obtained by first screening large numbers of a bacteria, *Salmonella typhimurium*, to identify resistant individuals. The coding sequence for the specific DNA responsible for the bacterial EPSPS synthase in the resistant individuals was then identified and copies of the gene produced. Other examples of single enzyme systems amenable to genetic engineering are the defense enzymes that plants produce in response to fungal invasion, such as chitinases and glucanases and the anti-oxidant enzymes catalase and peroxidase, that protect the plant against stress. The DNA responsible for these individual enzymes has been identified.

4. Preparation of a construct. The new gene requires a promoter before it is expressed. A construct is built which contains the new DNA, a promoter and a genetic marker. The promoter might be of bacterial, viral or plant origin. The most commonly used promoter, particularly in dicotyledonous plants is derived from the cauliflower mosaic virus, CaMV35SRNA. Genetic markers which confer resistance to the antibiotic, kanamycin, or to herbicides or are visual such as the expression of β-glucuronidase or luciferase are included so as to be able to screen for the transgenic character *in vitro* or confirm that a transformation has occurred.

5. Introduction of gene into the vector. The bacteria *A. tumefaciens* and the related species *A. rhizogenese* are natural vectors since they infect the plant through wounds in the stem and root respectively, and in the process transfer a portion of their plasmid DNA to the host genome. This T-DNA causes the plant to synthesize opines which are a nutrient source for the *A. tumefaciens* and also to form galls as a result of overproduction of auxin and cytokinin. The plasmids have been modified so that they still have the capacity to insert DNA into the plant genome but do not contain DNA for enhanced auxin and cytokinin expression or for opine synthesis. The DNA construct containing promoter, genetic marker and foreign gene can be inserted into this modified plasmid. The *A. tumefaciens* do not infect monocots very readily.

6. Transfer of gene into the host plant. Transformation requires that the gene is introduced into the cell genome through cell wall, plasmalemma, cytoplasm and nuclear envelope without loss of function. This transfer either relies on the transfer system of the natural DNA transfer of the *A. tumefaciens*, or by other mechanical means. These latter methods make use of a microprojectile bombardment, where small tungsten balls are coated in the construct DNA then fired into the cells, or alternatively, sharp DNA coated silicon carbide fibers have been used to mechanically penetrate the cells and introduce the DNA. In a more complicated process, the barrier of the cell wall is removed and the protoplast is able to take up DNA in the presence of polyethylene glycol (PEG), or is stimulated by electroporation or the DNA inserted by microinjection. The mechanical methods are more suitable for monocots.

7. Select transformed cells. The presence of a genetic marker means that the transformed cells or organ cultures can be incubated on a medium *in vitro* containing kanamycin, or a herbicide, and only those transformed cells or tissues will survive. The cells and tissues need to be regenerated to plants then exposed to a further screen on antibiotic or herbicide to confirm the transformation.

8. Test for transformation. Final confirmation of the transformed plants is by identification of the inserted genetic sequence in the host plant genome.

9. Stability of transformant. Field testing of the transformed plants is undertaken to ensure that the characters such as herbicide resistance, or disease resistance are expressed under field conditions and are stable through progeny generations.

The above strategy is the same for a wider range of species, more specific protocols can be obtained from the literature.

14.2 Gene manipulation with *A. tumefaciens*

The advantage of *A. tumefaciens* over other transformation techniques is that it is possible to generate the transformed cells at relatively high frequency and with no significant decline in regeneration capacity. Since a specific portion of DNA is transferred to the plant genome via the T-DNA, it means that the foreign DNA is defined and is inserted in the genome intact with no major rearrangements. The basic process of co-incubating callus or explant tissue with the *A. tumefaciens*, followed by regeneration of the transformed tissue is relatively straightforward. In the past, the limit to its general use was the capacity to infect a wide range of species and then to be able to regenerate the transformed tissue. The situation now is that it is possible to transform many species, including, quite recently, the monocots, maize and rice, and to regenerate the transformants.

14.2.1 Tobacco

The protocol for tobacco has been presented (K.P. Croft, personal communication) since it is a species that is widely used as a model system and also grown as a crop.

1. Young leaves, approximately 500 mg fresh weight, are taken from the upper leaves of a 4 week old tobacco plant grown *in vitro* (see Chapter 7) and are cut into 1 cm squares, avoiding the midrib.

2. A culture of *A. tumefaciens* containing a construct of the promoter, CaMV35SRNA, the antibiotic-resistant gene for kanamycin resistance and the foreign gene is streaked on LA medium (1% bactotryptone, 0.5% yeast extract, 1% NaCl and 1 g l⁻¹ agar) in petri dishes containing 50 mg l⁻¹ kanamycin. After overnight growth at 28°C, a single colony is inoculated into LB medium (1% bactotryptone, 0.5% yeast extract, 1% NaCl) containing 50 mg l⁻¹ kanamycin and incubated overnight at 28°C. The culture is centrifuged at 100g for 10 min and resuspended in 100 ml incubation media (MS media and 3% sucrose, pH 5.9), then 20 ml of this solution is distributed to five 9 cm petri dishes.

3. The tobacco leaves are placed in the *A.tumefaciens* solution for 20 min after which they are placed on plates containing callus initiation media (MS media, 1 mg l⁻¹ BAP, 0.1 mg l⁻¹ NAA , 3% sucrose, 0.8% bactoagar, pH 5.9) for 24 h in a plant growth room at 25°C.

4. The explants are transferred to glass or clear plastic containers onto a selective medium which is the same as in step 3, but which includes 100 mg l⁻¹ kanamycin to select for transformed cells, and 500 mg l⁻¹ carbenicillin to kill the *A.tumefaciens*. After 6 weeks incubation in the light at 25°C, shoots will appear on the callus tissue.

5. Shoots are transferred into fresh containers onto a rooting medium (0.5 MS medium) with 100 mg l⁻¹ kanamycin and 200 mg

l^{-1} carbenicillin. Those shoots that are rooted within 3 weeks are transferred to fresh containers for a further period of growth.

6. Rooted shoots are transferred to soil in 13 cm pots. For the first 2–3 days a large plastic cover is placed over the plants to reduce excessive transpiration and to encourage the development of a cuticle (see Chapter 13).

7. The transformation is confirmed by allowing the plants to flower with the heads bagged to encourage selfing. The selfed seeds are surface-sterilized by incubating in 10% Domestos for 5 min, followed by washing with sterile water five times. The seeds are then placed on a selective medium (0.5 MS and 2% sucrose, 0.8% bactoagar) containing 100 mg l^{-1} kanamycin in 9 cm petri dishes and incubated under low light at 25°C for 3 weeks. Yellow seedlings are sensitive to kanamycin and do not carry the T-DNA whereas those seedlings which remain green and grow strongly are transformed.

14.2.2 Potato

The potato is a good example of an important crop species which can be transformed by *A. tumefaciens*. The principle is similar to the tobacco [1].

1. Shoot tip cultures of the potato are initiated to provide a constant source of surface-sterile tissue. Axillary buds are sterilized in 2% Na hypochlorite plus 0.1% Tween (a commercial detergent), followed by 3–5 washes in sterile water, and are then placed on a growth medium (MS medium, 1% sucrose, 8 g l^{-1} agar) in the light at 25°C and subcultured every 3–4 weeks.

2. The tissue is pre-cultured by placing cut leaves for 1–3 days on petri dishes containing MS medium, 30 g l^{-1} sucrose, 8 g l^{-1} agar which is overlayed with 2 ml MS medium, 3% sucrose, 0.5 mg l^{-1} thiamine-HCl and 0.5 mg l^{-1} pyridoxine-HCl then covered with a piece of sterile Whatman No.1 filter paper at 25°C in the light.

3. A culture of *A. tumefaciens* is grown 2 days before the transformation in LB medium (1% bactotryptone, 0.5% yeast extract, 1% NaCl) then overnight, the culture is incubated with the selectable antibiotic, kanamycin (50 mg l^{-1}). The overnight culture is diluted 1:10 in liquid medium (MS medium, 1% sucrose) then the pre-cultured explants are immersed for 10 min in the diluted culture of *A. tumefaciens* culture, blotted dry then

returned to the pretreatment petri dishes and incubated at low light intensity for 1–2 days. After this time the explants are placed on a medium (MS medium, 10 g l^{-1} sucrose, 2.0 mg l^{-1} zeatin, 0.01 mg l^{-1} NAA, 0.1 mg l^{-1} GA$_3$) containing antibiotics cefotaxime or carbenicellin to remove the *A. tumefaciens*. The explants after 3–5 days are transferred to the selection medium. This medium (MS medium, 1% sucrose, 2.0 mg l^{-1} zeatin, 0.01 mg l^{-1} NAA, 0.1 mg l^{-1} GA$_3$, 8.0 g l^{-1} agar) contains cefotaxine, 200 mg l^{-1}, as before and the selectable antibiotic, kanamycin, 50 mg l^{-1}). The explants are transferred every 2 weeks to fresh medium.

4. Green callus are removed after 3–6 weeks and transferred to plates containing a shoot induction medium (MS medium, 10 g l^{-1} sucrose, 0.25 mg l^{-1} BA, 0.1 mg l^{-1} GA$_3$, 200 mg l^{-1} cefotaxime, 50 mg l^{-1} kanamycin, 8 g l^{-1} agar). Those callus with shoots are transferred to the selection medium with antibiotics (MS medium 30 g l^{-1} sucrose 200 mg l^{-1} cefotaxime, 50 mg l^{-1} kanamycin) then rooted shoots are transferred to fresh selection medium of the same composition.

5. When small roots appear, the agar is removed from the base of the shoot and the rooted shoots transferred to sterile soil in a humidity chamber to harden off. After 14 days the plants are transferred to larger pots and grown under standard growth conditions.

14.3 Non-tissue culture based transformation using *A. tumefaciens*

The disadvantage of the method of infection described for tobacco and potato is that the plant spends some time as an undifferentiated tissue culture. There is always the risk that somaclonal variation is induced during this period of culture. The problem could be avoided and the process simplified by eliminating the tissue culture stage. In a new approach, intact plant tissue is co-incubated with a culture of *A. tumefaciens* containing an appropriate construct. In one method the *A. tumefaciens* culture is co-incubated with seeds, when the *A. tumefaciens* is thought to remain in the seedling and adult plant without infecting the cells, but then enter and transform the zygote. The approach has been further modified so that now the flowering head is submerged in the culture of *A. tumefaciens* culture, and the bacteria are introduced into the plant by vacuum filtration. Treated plants are allowed to flower and set seed then the seed germinated and the transformed plants identified by screening the seedlings for resistance to an antibiotic-resistant gene that was part of the original T-DNA. Besides eliminating the risk of somaclonal variation, this method is relatively simple since there is no need for a succession of media to induce regeneration of shoots and roots. So far this approach has not been applied to a large number of species.

14.4 Tissue culture transformation not based on *A. tumefaciens*

There is much concern over the possible environmental dangers of genetically engineered plants and the use of *A. tumefaciens* comes in for some criticism. The *A. tumefaciens* infection seems to remain on and in transformed plants which means that the genetically modified *A. tumefaciens* is available as a pathogen for other crop or weed species. It is possible that herbicide resistance for instance could be introduced into weeds in this way. The weeds would then become as resistant to the herbicide as the crop. There are a number of methods that do not involve *A. tumefaciens*.

14.4.1 Uptake of DNA into protoplasts

Development of protoplast methodology has provided a material that is most suitable for transformation. The cells are single cells, so following transformation they will form cultures and plants without the danger of chimeras. In addition the absence of a cell wall makes it much easier to introduce large molecules directly through the plasmalemma. Since the monocots have proved largely resistant to infection by *A. tumefaciens*, alternative methods have been developed for this group but the methods are equally applicable to dicots. These methods rely on the introduction of DNA through the plasmalemma of the protoplasts, then regeneration of the protoplasts. Success at the regeneration stage has been a critical requirement but now protoplast regeneration has been achieved in most cereal protoplasts by using embryogenic cell suspensions as source material for protoplast isolation [2].

Uptake of the DNA has been through co-incubation of the protoplasts and gene construct in the presence of PEG, or alternatively by the use of electroporation where the protoplasts are exposed briefly to a short pulse of direct current. Both the PEG treatment and the electroporation have the effect of causing the membrane to become sticky so that it will enclose any large molecule or plasmid. Because electroporation appears to cause less cell damage and is relatively simple and inexpensive, this technique is favored rather than exposure of the cells to PEG.

14.4.2 Microprojectile bombardment of plant tissue

Because it is difficult to infect monocots with *A. tumefaciens,* the alternative for many monocots has been to use particle bombardment. This involves firing tungsten microprojectiles (4 μm diameter) covered in the DNA construct into tissue. This construct would have a promoter, foreign gene and marker gene as before. The plant tissue that is most suitable for use is one that has retained its potential for regeneration, for example, immature embryos, seed-derived callus, and proliferative shoot cultures. In maize, for example, a suspension culture of embryogenic cells is bombarded with particles followed by selection for transformed cells by screening in the presence of a herbicide. Resistance to the herbicide would be encoded as a marker gene in the original DNA construct. In rice, immature embryos, embryogenic callus, immature inflorescences and mature seed-derived callus and in wheat, precultured scutellum of immature embryos provides the source tissue for bombardment.

The advantage of the microprojectile system is the reduction in time that the material spends as callus or cell suspensions, so that the risks of somaclonal variation are much lower. In some instances, the tissue culture stage has been eliminated altogether as meristematic parts of intact plants have been the focus of bombardment. The loss of the tissue culture stage also means a great saving in time and facilities in any plant breeding excercise. The other major advantage is the high transformation frequency which for rice has been estimated as 50% of the number of explants bombarded.

Although the bombardment method has been very successful for recalcitrant species, it does have its drawbacks. At present, the details of the integration process are not well known, so it is not possible to improve it from fundamental principles. The other drawback is the high cost of the equipment. This means that well-funded laboratories will have this equipment but the many laboratories in the developing world who wish to exploit this area will not be able to. The fact that rice can now be transformed using *A. tumefaciens* does help to remove this financial restriction.

14.4.3 Wounding of plant tissue by silicon carbide fibers

A further development which is both simple and effective is the use of silicon carbide fibers [3]. These are single crystrals with an average diameter of 0.6 µm and length ranging from 10–80 µm and a high tensile strength. The starting plant material is a cell suspension which has retained the capacity to regenerate. It is mixed by rapid vortexing with a suspension of the fibers and plasmid DNA. The sharp ends of the fibers wound the cell minutely and allow the plasmid to be taken up into the cell cytoplasm, as if it were introduced by *A. tumefaciens*. Electron microscope examination of the cell showed that fibers appear to penetrate the cell wall and cytoplasm. The DNA is thought to enter the cell by adhering to the surface of the fibers. Transformation frequencies have been similar to those achieved with bombardment methods.

The biggest problem with this method is the limited penetration in potentially regenerating systems, such as embryonic cell suspensions of monocots. In these cultures the cell structure may be altered so that the cells are much less flexible and have thicker cell walls, thus limiting entry by the fibers. The parameters that need to be considered are genotype and species differences and practical issues such as the fiber DNA-to-plant-cell ratio, vortexing method and methods ensuring the survival of the transformed cells. The primary disadvantage is the risk of lung damage and the carcinogenic properties of the fibers if inhaled. It is important that both the dry material and suspensions are handled in exhaust hoods.

14.5 Genetically engineered characters in crop plants

Historically, and for strictly commercial reasons, the greatest effort in plant genetic engineering has been in the development of herbicide resistance in plants. Herbicide resistance to glyphosate, for example, has now been achieved and despite some public concern, this resistance is being introduced into a range of crop species including soybean and oil seed rape. Other characters of interest to the plant breeder and food technologist, are the improvements in resistance to disease, insect pests, environmental stress and in more specialized characteristics, such as the control of ripening of fruit since this affects their shelf life and therefore their appearance and attractiveness [4].

Plants have been transformed by enhancing the level of the defense enzymes chitinase and β-1-3-glucanase so as to increase the resistance to fungal infection. Normally these enzymes break down the cell wall of the invading fungus and either kill the fungus or restrict its proliferation. Chitinase and β-1-3 glucanase enzymes are only produced by the plant as an induced response to the presence of the fungus in the tissue, but by that stage the crop may be visually damaged and the yield reduced. By engineering a constitutive and more active expression of the two defense enzymes in the crop plant, there is no delay before the fungus is attacked, so that levels of the pathogen do not build up.

Another example of single gene characters are those involved in the defense of plants against insect attack. The bacterium, *Bacillus thuringiensis* var kurstaki, produces a protein in the spores which have been used as a insecticidal spray for 30 years. When the plant with the spores on its surface is eaten by an insect, the protein becomes cleaved in the insect gut and the toxin released permeabilizes the cells that line the gut and reduces their digestive function. Now the DNA for this protein has been modified to increase the toxicity of the protein and inserted into cotton, tomato and tobacco plants. The protein has been expressed with some insecticidal effect against the primary lepidopteran pests. A further example of protection agaist insect attack is that provided by a trypsin inhibitor which slows down digestion of proteins by the insect. This protein has been identified in one species, the cow pea and transferred to another species, tobacco, where it increase resistance to the tobacco bud worm.

Resistance to fungal and microbial attack is also dependent on the production phytoalexins following invasion. These compounds, which are secondary products have complex pathways which are only now being understood (see Chapter 12). The key enzymes in these pathways are being identified and their activity enhanced in crop plants by the insertion of multiple copies of the gene for the control enzyme.

Resistance to environmental stresses such as drought, low or high temperature, salinity and heavy metals is thought to be polygenic, but every effort is now being made to identify key enzymes in the resistance response. For example, plants under stress are associated with the production of free radicals. These are destructive to cell membranes particularly and will ultimately damage the plant. The resistant response involves the production of anti-oxidant enzymes such as peroxidases and catalases which remove the free radicals and thus reduce the potential for damage. The presence of anti-oxidant enzymes provides a route by which the resistance of the plants can be modified by gene amplification of single protein systems.

References

1. **Visser, R.G.F.** (1991) In *Plant Tissue Culture Manual – Fundamentals and Application* (ed. K. Lindey). Kluwer Academic Publishers, The Netherlands, **B5**, 1–9.
2. **Kyozuka, J. and Shimamoto, K.** (1991) *Ibid,* **B2**, 1–17.
3. **Frame, B.R., Drayton, P.R. Bagnall, S.V., Lewnau, C.J., Bullock, W.P., Wilson, H.M., Dunwell, J.M., Thompson, J.A. and Wang K.** (1995) In *Current Issues in Plant Molecular and Cellular Biology* (eds M. Terzi, R. Cella and A. Falavigna). Kluwer Academic Publishers, The Netherlands, pp. 279–284.
4. **Shah, D.M., Rommens, C.M.T. and Beachy, R.N.** (1995) *Trends Biotechnol.* **13**, 362–368.

Appendix A

Sources and suppliers for chemicals and equipment

Tissue culture media

Complete media and separate components

Sigma Chemical Company, Fancy Road, Poole, Dorset BH17 7NH, UK.

ICN Pharmaceuticals Inc., 3300 Hyland Avenue, Costa Mesa, CA 92626, USA.

ICN Biochemicals Ltd, Unit 18 Thame Park, Business Centre, Wenman Road, Thame, Oxfordshire OX9 3XA, UK.

Glass and plastic ware

Conical flasks, Universal vials, centrifuge tubes (screw-capped)

Merck BDH, Merck Ltd, Hunter Boulevard, Magna Park, Lutterworth, Leicestershire LE7 4XN UK.

Philip Harris, Lynn Lane, Shenstone, Lichfield, Staffordshire WS14 0EE, UK.

Sterile plastic dishes, 9 cm

Sterilin Ltd, 43/45 Broad Street, Teddington, Middlesex TW11 8QZ, UK.

Glass petri dishes, 5 cm

A/S Nunc, Kamstrupvej 99, Kamstrup, DK-4000 Roskilde, Denmark.

Gibco Bio-Cult Ltd. 3, Washington Road, Paisley PA3 4EP, UK.

Plastic (Magenta) containers | Sigma Chemical Company, Fancy Road, Poole, Dorset BH17 7NH, UK.

Aseptic facilities

Laminar flow cabinets | Philip Harris, Lynn Lane, Shenstone Lichfield, Staffordshire WS14 0EE, UK.

Antibiotics

Ampicillin | Boots the Chemist, UK.

Gentamycin, Tetracycline | Sigma Chemical Company, Fancy Road, Ltd, Poole, Dorset BH17 7NH, UK.

Carbencillin | Beecham Research Laboratories, Brentford, Middlesex, UK.

Enzymes used in protoplast isolation

Cellulase
Cellulysin | Calbiochem, La Jolla, CA, USA.

Cellulase
Hemicellulase | Sigma Chemical Company, Fancy Road, Poole, Dorset BH17 4EP, UK.

Meicelase | Meiji Seika Kaisha, Tokyo, Japan.
Macerozyme R10 | Kinki Yakult, Nishinomiya, Japan.

Genetic manipulation

Biolistic appparatus and electroporation | BioRad Laboratories, 200 Alfred Noble Drive, Hercules, CA 94547, USA.

BioRad Laboratories Ltd, BioRad House, Marylands Avenue, Hemel Hempstead, Hertfordshire HP2 7TD, UK.

Appendix B

Further reading

Dixon, R.A. (1985) *Plant Cell Culture – A Practical Approach.* IRL Press, Oxford, UK.

Dixon, R.A. and Gonzales, R.A. (1994) *Plant Cell Culture – A Practical Approach.* IRL Press, Oxford, UK.

Evans, D.A, Sharp, Ammirato, Yamada,Y. (eds) (1983) Part A: Basic techniques of plant cell culture. In *Handbook of Plant Cell Culture – Volume 1, Techniques of Plant Cell Culture.* Macmillan Publishing Co., New York and Collier Macmillan Publishers, London.

Goddijn, O.J.M. and Pen, J. (1995) Plants as bioreactors. *Trends Biotechnol.* 13, 379–387.

Harper, P.C. (1991) Micropropagation in bulb crops. In *Horticulture – New Technologies and Applications. Current Plant Science and Biotechnology in Agriculture* (J. Prakash and R.L.M. Pierik, eds). Kluwer Academic Publishers, London, UK, pp. 161–168.

Hughes, M.A. (1996) *Plant Molecular Genetics.* Longman, Harlow, UK.

Hunter, C.F. (1993) *In Vitro Cultivation of Plant Cells.* Butterworth Heinemann, Oxford, UK.

Lindsey, L. (ed.) (1991) *Plant Tissue Culture Manual – Fundamentals and Applications.* Kluwer Academic Publishers, The Netherlands.

Potrykus, I. (1991) Gene transfer to plants – Assessment of published approaches and results. *Annu. Rev. Plant Physiol. Plant Mol. Biol.* **42,** 205–225.

Reinert, J. and Bajaj, Y.P.S. (eds) (1977) *Applied and Fundamental Aspects of Plant Cell, Tissue and Organ Culture.* Springer-Verlag, Berlin, Germany.

Stafford, A. and Warren, G. (eds) (1991) *Plant Cell and Tissue Culture.* Open University Press, Milton Keynes, UK.

Walden, R. and Wingender, R. (1995) Gene transfer and plant regeneration techniques. *Trends Plant Technol.* **13**, 324–331.

Index